Beach Nourishment

Beach Nourishment: A 21st Century Review

Editors

Juan J. Munoz-Perez
Luis Moreno

MDPI • Basel • Beijing • Wuhan • Barcelona • Belgrade • Manchester • Tokyo • Cluj • Tianjin

Editors
Juan J. Munoz-Perez
Applied Physics
University of Cadiz
Puerto Real
Spain

Luis Moreno
Hydraulics, Energy and
Environmental Engineering
Politecnica de Madrid
Madrid
Spain

Editorial Office
MDPI
St. Alban-Anlage 66
4052 Basel, Switzerland

This is a reprint of articles from the Special Issue published online in the open access journal *Journal of Marine Science and Engineering* (ISSN 2077-1312) (available at: www.mdpi.com/journal/jmse/special_issues/beach_nourishment).

For citation purposes, cite each article independently as indicated on the article page online and as indicated below:

LastName, A.A.; LastName, B.B.; LastName, C.C. Article Title. *Journal Name* **Year**, *Volume Number*, Page Range.

ISBN 978-3-0365-1606-6 (Hbk)
ISBN 978-3-0365-1605-9 (PDF)

Contents

About the Editors

Juan J. Munoz-Perez

Dr. Juan J. Muñoz-Perez has worked in private companies both in consulting and public works, as well as in the ports of Barcelona and Cadiz. As Chief Engineer of the Atlantic Coastal Department of Andalusia (Spain) has planned, supervised, and directed nearly 400 maritime works. He has combined the former tasks with teaching as Professor at the University of Cadiz since 1991 for the grades of Sea Sciences and Civil Engineering. He is the author of numerous contributions in congresses and articles in prestigious journals as a result of the projects he has directed through the Coastal Engineering research group that he runs (scholar.google.es/citations?user=xyE3rnoAAAAJ&hl=es). He currently manages the Campus of International Excellence of the Sea or CEIMAR (https://campusdelmar.com/)

Luis Moreno

Dr. Luis J. Moreno has more than 30 years of professional experience in the field of Coastal and Oceanographic Engineering consultancy in a variety of roles, not only in the public sector but also in the private sector. Has successfully managed a wide range of projects in the Middle East, Europe, Latin America, and Africa.

Dr. Moreno is a recognized professional who regularly participates in technical conferences and meetings, as well as in national and international postgraduate and master's courses. He served as a national delegate at the Committee of Advisory Nature/Marine Science and Technology of the European Union and has been a project evaluator of R+D+i projects for the European Union.

As a part-time Associate Professor at Technical University of Madrid (UPM) since 2003, he has educated hundreds of students in harbor and coastal engineering as well as in port operation and infrastructure management.

Journal of
Marine Science and Engineering

MDPI

Editorial

Beach Nourishment: A 21st Century Review

Luis J. Moreno [1,*] and Juan J. Muñoz-Perez [2,*]

1 Department of Civil Engineering: Hydraulics, Energy and Environment, Technical University of Madrid, Prof. Aranguren 3, 28040 Madrid, Spain
2 CASEM, University of Cadiz, Rio San Pedro s/n, 11510 Puerto Real, Spain
* Correspondence: luisjuan.moreno@upm.es (L.J.M.); juanjose.munoz@uca.es (J.J.M.-P.)

Citation: Moreno, L.J.; Muñoz-Perez, J.J. Beach Nourishment: A 21st Century Review. *J. Mar. Sci. Eng.* **2021**, *9*, 499. https://doi.org/10.3390/jmse9050499

Received: 21 April 2021
Accepted: 29 April 2021
Published: 5 May 2021

Long-term erosion is experienced by most of the coastlines worldwide, and it is usually attributed not only to sea level rise but also to the retention of sand in dams, the occupation of dry beaches by urbanized areas, the disturbance of the natural patterns of longshore drift, the mining of sand as building material for construction, and so on. Beach nourishment has evolved as the favored erosion-mitigation strategy in many areas of the world. The increasing number of people living on the coast, the safety of those people, and the high values of coastal property [1] are all factors that have made beach nourishment a cost-effective strategy for managing erosion in many locations. However, a new scenario of sand scarcity and environmental care has arisen in recent decades [2]. There have been a number of different and interesting cases of various aspects of beach nourishment in the last years. The purpose of this Special Issue has been to publish the different experiences and research related to this topic.

After a careful review process, nine papers were included. Their thematic contributions include the use of field methods such as the use of remotely piloted aircraft systems (RPAS) or un-manned aerial vehicles (UAV) for faster and automated mapping of the coastal area or the acquisition of geomagnetic data in marine environments; the use of multi-approach methodologies to assess the interaction between coastal structures and beaches and in particular of submerged pipelines; the need to adopt a plan for the optimal use of limited resources of available sediment from a regional perspective and the assessment of the effectiveness of beach nourishments; the understanding of the role of submerged geological control of beach profiles together with the implementation of innovative beach nourishment strategies while facing the non-trivial challenge of visualizing and communicating mesoscale modeling assumptions, uncertainties and outcomes to both coastal specialists and decision makers; and the influence of sea-level rise and erosion on diminution of beach habitats.

The contributions are commented upon in order of appearance in this Special Issue. Although an effort has been made to compile contributions that cover an update in the state-of-the-art of innovative techniques in beach nourishment, by no means should they be limited to the topics presented hereby.

To begin with, the size and great dynamism of coastal systems require faster and more automated mapping methods, such as the use of a remotely piloted aircraft system (RPAS) or unmanned aerial vehicle (UAV). However, the main problem for surveying using low-altitude digital photogrammetry in beach areas is their visual homogeneity. Obviously, the fewer the homologous points defined by the software, the lower the accuracy. Contreras-de-Villar et al. [3] have addressed the error performed in photogrammetric techniques, such as flight height, flight time, percentage of frame overlap (side and forward), and the number of ground control points (GCPs). Among their conclusions, it should be highlighted that the error for noon flights is almost double that for early morning flights. Moreover, a minimum value of 7 GCP per hectare should be taken into account when designing a beach leveling campaign using RPAS.

Coastal areas are usually very impacted because of demographic and industrial pressure, which leads to an interaction between anthropic infrastructures and littoral dynamics. One of the public works that can most influence the sediment transport is a submerged pipeline. Lersundi-Kanpistegi et al. [4] studied the extension of the wastewater pipeline in Vigo (Northwest Spain) crossing the most important urban beach of the city by using a multi-tool strategy based on high resolution bathymetry data, seabed physical characterization, a grain size study of the superficial sediment, and a numerical simulation of the tide, wave climate, and sediment transport in low and high energy conditions using the open source Delft3D software. The results indicate characteristics that the design must follow in order to ensure that the future structure would not alter the global sediment dynamics of the beach. The multi-approach methodology presented can be applied to other studies of the interaction between coastal structures and beaches.

Beach nourishment is generally seen as the preferred means of rectifying coastal erosion, due to its low environmental impact and natural evolution. Martell et al. [5] present a study regarding the effectiveness of beach nourishments in Cancun (Mexico), but its conclusions regarding the erosion tendency directly linked to the incidence of extreme hydrodynamic conditions and the scarcity of natural sediment sources can be applied to beaches with similar characteristics in any other area. Furthermore, the need for improving long term predictions of the wave climate under global warming scenarios must be highlighted.

Submerged geological control of beach profiles, e.g., through the existence of reef flats or submerged sills, is a topic that has been widely studied over the last years. Moreover, fringing reefs have significant impacts on beach dynamics, yet there is little research on how they should be considered in beach nourishment design, monitoring, and conservation work. Thus, the behavior and characteristics of nourishment projects at two reef protected beaches, in Hawaii (USA) and in Cadiz (Spain), are compared [6] to provide transferable information for future nourishment projects and monitoring in this type of environment. Several differences were detected related to the nourishment cost, distance to the borrow site, post-nourishment monitoring frequency and assessment of accuracy, measurement of the beach volume increase after nourishment, etc.

Innovative beach nourishment strategies have been developed in the last decade, driven by the increased worldwide interest in environmentally friendly coastal protection measures. In this context, the massive nourishment project of the Netherlands (known as Sand Engine [7]) has been hailed as a successful means of beach protection. Adapting this idea, a very small and bell-shaped Sand Engine was designed to protect the beachfront at a tourist resort near Puerto Morelos, Mexico [8]. This micro Sand Engine is seen as a sustainable and eco-friendly coastal protection measure, especially applicable when large nourishment projects are not viable. Maintenance work for this type of nourishment is cost- and time-effective, and any negative impacts on sensitive ecosystems nearby can be detected and controlled quickly.

Coastal geomorphologists and engineers worldwide are increasingly facing the non-trivial challenge of visualizing and communicating mesoscale modeling assumptions, uncertainties and outcomes to both coastal specialists and decision makers. Payo et al. [9] show how the risk of simulation model outcomes can be minimized by using the Coastal Modeling Environment (CoastalME). CoastalME is a modeling framework for coastal mesoscale morphological modeling that can achieve close linkages between the scientific model abstractions and the 3D representation of topographic and bathymetric surfaces. A transparent methodology to merge the required variety of data types and formats into a 3D-thickness model is presented through the case study of Happisburgh (eastern England, UK). Finally, some of the barriers to the adoption of this methodology are analyzed.

Sometimes, the limited resources of available sediment make it necessary to adopt a plan for their optimal use from a regional perspective. This is the case presented by Pranzini et al. [10], who present a study carried out to support the Region of Tuscany Coastal Sediment Management Plan along the 215 km-long continental sandy coast of this

Italian region. Sand stability and color compatibility were determined in order to assess the possibility of using the available sediment in accreting sectors to nourish the beach in eroding areas. This kind of study is of great interest for the proposal of sound management actions to counteract the increasing erosion processes linked to climate change phenomena and human effects on rivers and coastal systems.

A method for the acquisition of geomagnetic data in marine environments, developed by the Oceanographic and Hydrographic Research Center of Colombia, is presented by Oviedo et al. [11]. Leaving sub-bottom profiling and side-scan sonar techniques aside, the most representative uses of the geomagnetic method are the location of pipelines and metal plates, detection of buried ordnance, identification of sites of archaeological interest, and the characterization of geological structures. To test the method, a grid of geomagnetic data was surveyed in an area close to the island of San Andrés (Northwest Colombian maritime territory) and compared with survey data obtained from National Oceanic and Atmospheric Administration (NOAA) magnetic data. Despite the long time interval between the two surveys, almost 50 years, no significant differences were observed in terms of the analyzed variables.

Finally, the influence of sea-level rise and erosion (along with shoreline hardening and reduced sediment inputs) on diminution of beach habitats is shown by Martin et al. [12]. Their study shows that increasing sandy beach habitat can be beneficial to wildlife, but the method of placement, timing of the project, and fate of the beach afterward can modulate or prevent beneficial effects. Frequent repetition of sand placement may accumulate impacts without allowing sufficient time for the ecosystem to recover.

Closing this editorial, the guest editors consider that this Special Issue will provide benefits to technicians, engineers, researchers and managers in the area of beach nourishment.

Author Contributions: L.J.M. and J.J.M.-P. wrote and reviewed this editorial, and both authors have agreed to the published version of the manuscript. All authors have read and agreed to the published version of the manuscript.

Funding: This research received no external funding.

Acknowledgments: The Guest Editors wish to express their deepest appreciation to the authors for their cooperation, to the anonymous reviewers for their valuable comments, and to the Editor-in-Chief for the valuable comments, as well as to Esme Wang for her guidance, continuous support, and constructive advice throughout the publication process.

Conflicts of Interest: The authors declare no conflict of interest.

References

1. Houston, J.R. The economic value of America's beaches—A 2018 update. *Shore Beach* **2018**, *86*, 3–13.
2. Herrera, A.; Gomez-Pina, G.; Fages, L.; De La Casa, A.; Munoz-Perez, J.J. Environmental Impact of Beach Nourishment: A Case Study of the Rio San Pedro Beach (SW Spain). *Open Oceanogr. J.* **2010**, *4*, 32–41. [CrossRef]
3. Contreras-De-Villar, F.; García, F.J.; Muñoz-Perez, J.J.; Contreras-De-Villar, A.; Ruiz-Ortiz, V.; Lopez, P.; Garcia-López, S.; Jigena, B. Beach Leveling Using a Remotely Piloted Aircraft System (RPAS): Problems and Solutions. *J. Mar. Sci. Eng.* **2020**, *9*, 19. [CrossRef]
4. Bernabeu, A.M.; Lersundi-Kanpistegi, A.; Rey, D.; Díaz, R. Multidata Study to Evaluate the Impact of Submarine Outfall in a Beach Sedimentary Dynamic: The Case of Samil Beach (Galicia, Spain). *J. Mar. Sci. Eng.* **2020**, *8*, 461. [CrossRef]
5. Martell, R.; Mendoza, E.; Mariño-Tapia, I.; Odériz, I.; Silva, R. How Effective Were the Beach Nourishments at Cancun? *J. Mar. Sci. Eng.* **2020**, *8*, 388. [CrossRef]
6. Muñoz-Perez, J.J.; Gallop, S.L.; Moreno, L.J. A Comparison of Beach Nourishment Methodology and Performance at Two Fringing Reef Beaches in Waikiki (Hawaii, USA) and Cadiz (SW Spain). *J. Mar. Sci. Eng.* **2020**, *8*, 266. [CrossRef]
7. Stive, M.J.; De Schipper, M.A.; Luijendijk, A.P.; Aarninkhof, S.G.; Van Gelder-Maas, C.; Vries, J.S.V.T.D.; De Vries, S.; Henriquez, M.; Marx, S.; Ranasinghe, R. A New Alternative to Saving Our Beaches from Sea-Level Rise: The Sand Engine. *J. Coast. Res.* **2013**, *290*, 1001–1008. [CrossRef]
8. Escudero, M.; Mendoza, E.; Silva, R. Micro Sand Engine Beach Stabilization Strategy at Puerto Morelos, Mexico. *J. Mar. Sci. Eng.* **2020**, *8*, 247. [CrossRef]
9. Payo, A.; French, J.R.; Sutherland, J.; Ellis, M.A.; Walkden, M. Communicating Simulation Outputs of Mesoscale Coastal Evolution to Specialist and Non-Specialist Audiences. *J. Mar. Sci. Eng.* **2020**, *8*, 235. [CrossRef]

10. Pranzini, E.; Cinelli, I.; Cipriani, L.E.; Anfuso, G. An Integrated Coastal Sediment Management Plan: The Example of the Tuscany Region (Italy). *J. Mar. Sci. Eng.* **2020**, *8*, 33. [CrossRef]
11. Prada, K.O.; Antelo, B.J.; Murillo, N.O.; Cózar, J.R.; Contreras-De-Villar, F.; Muñoz-Pérez, J.J. A New Method for the Collection of Marine Geomagnetic Information: Survey Application in the Colombian Caribbean. *J. Mar. Sci. Eng.* **2020**, *9*, 10. [CrossRef]
12. Martin, K.L.M.; Adams, L.C. Effects of Repeated Sand Replenishment Projects on Runs of a Beach-Spawning Fish, the California Grunion. *J. Mar. Sci. Eng.* **2020**, *8*, 178. [CrossRef]

Journal of
Marine Science and Engineering

MDPI

Article

Beach Leveling Using a Remotely Piloted Aircraft System (RPAS): Problems and Solutions

Francisco Contreras-de-Villar [1,*], Francisco J. García [2], Juan J. Muñoz-Perez [1], Antonio Contreras-de-Villar [1], Veronica Ruiz-Ortiz [1], Patricia Lopez [1], Santiago Garcia-López [3] and Bismarck Jigena [4]

1 Department of Industrial Engineering and Civil Engineering, Campus Bay of Algeciras, University of Cadiz, Avda. Ramón Puyol s/n, 11202 Algeciras, Spain; juanjose.munoz@uca.es (J.J.M.-P.); antonio.contreras@uca.es (A.C.-d.-V.); veronica.ruiz@uca.es (V.R.-O.); patricia.lopezgarcia@uca.es (P.L.)
2 Urbing-Lab Diseño y Gestión, C/Cabo Noval 6, 52005 Melilla, Spain; urbing.lab@gmail.com
3 Department of Earth Sciences, Campus Río San Pedro s/n, University of Cadiz, 11510 Puerto Real, Spain; santiago.garcia@uca.es
4 Department of Nautical Sciences and Maritime Studies, Campus Río San Pedro s/n, University of Cadiz, 11510 Puerto Real, Spain; bismarck.jigena@uca.es
* Correspondence: francisco.contreras@uca.es

Abstract: The size and great dynamism of coastal systems require faster and more automated mapping methods like the use of a remotely piloted aircraft system (RPAS) or unmanned aerial vehicle (UAV). This method allows for shorter intervals between surveys. The main problem for surveying using low-altitude digital photogrammetry in beach areas is their visual homogeneity. Obviously, the fewer the homologous points defined by the program, the lower the accuracy. Moreover, some factors influence the error performed in photogrammetric techniques, such as flight height, flight time, percentage of frame overlap (side and forward), and the number of ground control points (GCPs). A total of 72 different cases were conducted varying these factors, and the results were analyzed. Among the conclusions, it should be highlighted that the error for noon flights is almost double that for the early morning flights. Secondly, there is no appreciable difference regarding the side overlap. But, on the other side, RMSE increased to three times (from 0.05 to 0.15 m) when forward overlap decreased from 85% to 70%. Moreover, relative accuracy is 0.05% of the flying height which means a significant increase in error (66%) between flights performed at 60 and 100 m height). Furthermore, the median of the error for noon flights (0.12 m) is almost double that for the early morning flights (0.07 m) because of the higher percentage of grids with data for early flights. Therefore, beach levelings must never be performed at noon when carried out by RPAS. Eventually, a new parameter has been considered: the relationship between the number of GCPs and the surface to be monitored. A minimum value of 7 GCP/Ha should be taken into account when designing a beach leveling campaign using RPAS.

Keywords: UAV; RPAS; littoral systems; aerial photogrammetry; DTM; monitoring; SfM; GCPs

Citation: Contreras-de-Villar, F.; García, F.J.; Muñoz-Perez, J.J.; Contreras-de-Villar, A.; Ruiz-Ortiz, V.; Lopez, P.; Garcia-López, S.; Jigena, B. Beach Leveling Using a Remotely Piloted Aircraft System (RPAS): Problems and Solutions. *J. Mar. Sci. Eng.* **2021**, *9*, 19. https://doi.org/10.3390/jmse9010019

Received: 23 November 2020
Accepted: 21 December 2020
Published: 26 December 2020

1. Introduction

Coastal erosion has become one of the most important concerns of different countries [1]. Coastal areas are the focal points of tourist attractions, which translates into an important source of economic income [2–4]. Moreover, the study of coastal behavior helps us understand the complex processes that occur in these areas [5,6]. Their understanding leads us to the prevention of coastal erosion, and thus monitoring the evolution of our beaches is essential [6]. Thus, a methodology for carrying out measurements of some oceanographic phenomena using Unmanned Aerial Vehicles (UAV also known as remotely piloted aircraft system or RPAS) have already been presented by other researchers (e.g., [7]). Nevertheless, some aspects can still be taken into account as we will show later.

Correct coastal modeling needs a three-dimensional reconstruction of the study area [8]. Coastal modeling is represented by digital terrain models (DTMs) of high spatial resolution. Geomorphological state can be defined as a multitemporal surface [9,10]. Depending on the beach area to be mapped (dry zone, intertidal zone, or submerged zone), various techniques and methodologies can be used [11,12]. The dry beach and the intertidal zone have been mapped using direct topography techniques [12]. Initially, a tachymeter was used, being later replaced by the total station, an electronic transit theodolite integrated with electronic distance measurement (EDM), and an on-board computer to collect data and perform triangulation calculations. This task is currently done with GPS techniques. This type of point-to-point data collection is cheaper than the previous ones because it only requires one technician. However, GPS surveying is limited to small and easily accessible areas.

The size and great dynamism of these coastal systems require faster and more automated mapping methods [13]. Thus, the synchronous nature of the data is not lost [14]. Photogrammetry has evolved to the technique called structure from motion (SfM) [13–17] based on algorithms that allows one to obtain excellent cartographic results from a set of frames that cover an area. The emergence of RPAS systems as well as the high definition of today's digital cameras have induced new cartographic systems. The use of these systems significantly reduces costs and execution times, providing excellent accuracy [18]. This technology, tested in multiple applications, appears as a serious competitor against other cartographic techniques (e.g., LIght Detection and Ranging or LIDAR) [19]. Hugengoltz et al. [20], for instance, stated that the vertical RMSE of an RPAS data set was equivalent to the RMSE of a bare earth LiDAR DTM for the same site.

The work procedure involves the definition of a series of parameters such as flight height (Figure 1a), covering area, percentage of overlap between adjacent frames (Figure 1b,c), and a different number of ground control points (GCPs) [21] (Figure 1d).

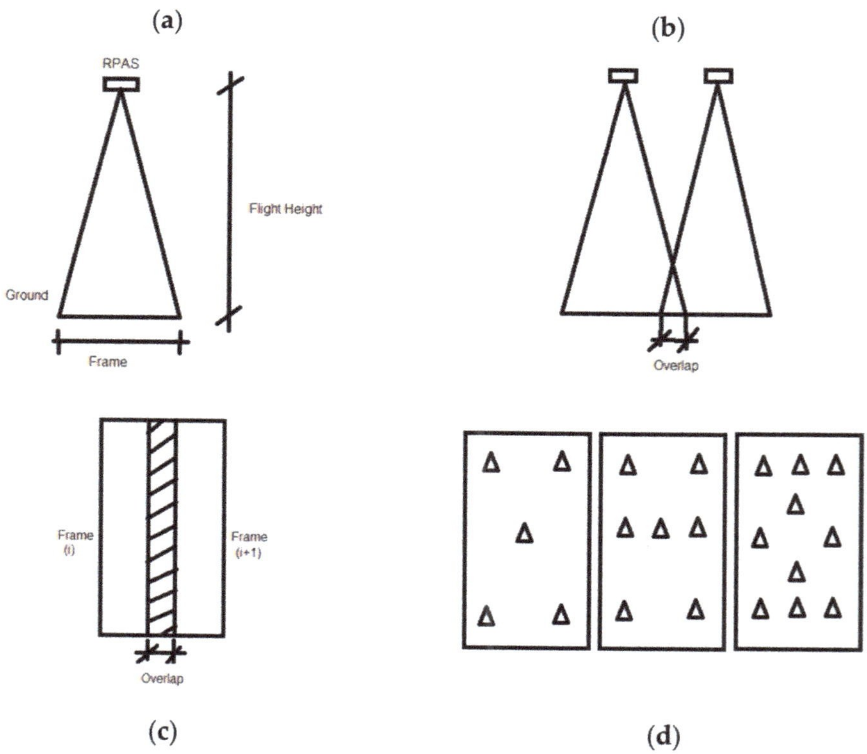

Figure 1. Sketch showing basic concepts of the remotely piloted aircraft system (RPAS) system: frame and flight height (**a**), overlap between two adjacent frames (**b**,**c**) and distribution of the different number of ground control points (GCPs) (**d**).

A particular case of the problems of low-altitude digital photogrammetry is the identification of common points in contiguous frames over poorly differentiated visual areas. When performing a low-altitude flight over highly homogeneous surfaces (beach sand, snow, agricultural areas of the same crop) is difficult to find common points [22]. The fewer the homologous points defined by the program, the lower the accuracy. This fact is common in the photogrammetry of beach areas and the accuracy of the DTM will be the result of the concatenation of the errors in different stages [23].

Thus, this paper aims to compare the vertical accuracy of a beach leveling, by using an RPAS, performed with different parameters of flight (height, time, side and forward overlap) and number of GCPs. Eventually, some guidelines will be presented to minimize the error of a photogrammetric survey.

2. Study Area

The chosen beach is Los Lances beach, in Tarifa (SW Spain). This area is considered as a bird special protection area due to its privileged situation that gives it a relevant role in air and marine migration processes (Figure 2).

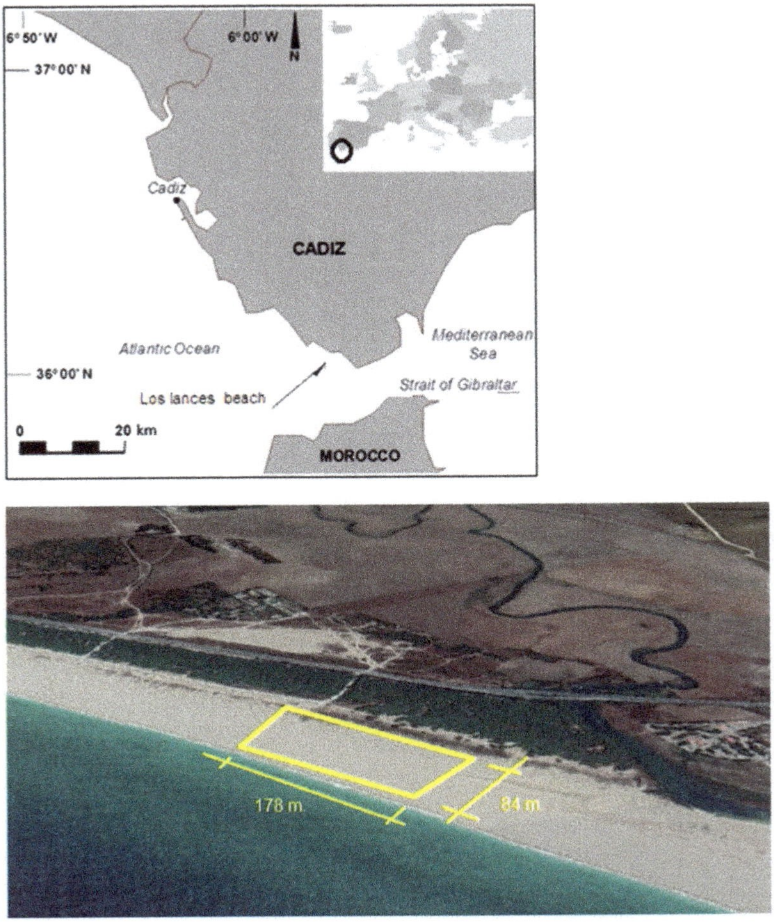

Figure 2. Location of the study area.

This space has a good state of conservation of ecosystems and a high-quality landscape. The beach is 3854 m long and covers an area of 280,000 m^2. The study area, a 178 m by 84 m rectangle, is also shown in Figure 2.

Its degree of urbanization is low. It has fine golden-colored sand, composed of medium-coarse unconsolidated sediments. The D_{50} of the emerged sand is 0.34 mm. It is a dissipative beach and has waves of medium-moderate degree [24]. The maximum tidal range is about 1.40 m, and the significant wave height Hs is about 3.7 m [21]. It is a type of semi-urban beach widely used by windsurfers and kitesurfers due to the abundant windy days of the year that occur in this area. The prevailing winds in the area are eastwards and westwards [25]. The dry beach area before the dune area is over 100 m wide, which is an optimal area for conducting the study.

3. Methods

The factors that influence the error performed in digital photogrammetric techniques are flight height, overlap (side and forward), and GCP number. Different tests were conducted varying these factors, and the results were analyzed. The main problem for surveying using photogrammetric methods in beach areas is their visual homogeneity. This effect reduces the number of homologous points among neighboring frames and thus prevents optimal correlation.

In normal conditions, it is recommended to make the flights in hours close to noon since the sun is in the highest position, generating few shadows [26]. However, in addition to the noon flights, other flights were carried out early in the morning when the sun was at a low altitude. These supplementary flights were performed trying to find out if the shadows produce enough differentiation in the terrain as to decrease the margin of error (Figure 3).

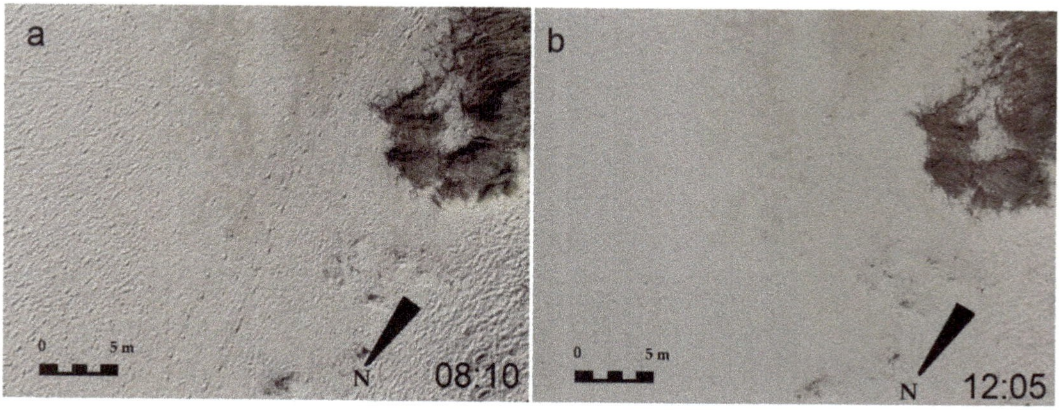

Figure 3. Frame detail: (**a**) early morning; (**b**) at noon. Frame details showing how shadows produce more homologous points in the early morning than at noon.

One difficulty is the short flight time of RPAS. Thus, the usual flight time (20–25 min) must be balanced against each other parameter: the surface to be flown, the flight height, and the overlap in the images we want to obtain.

The parameters that varied on each flight were the following:

- Data collection at 8 a.m. and 12 p.m.,
- Flight height at 60, 80, and 100 m,
- Side overlap at 85% and 70%,
- Forward overlap at 85% and 70%,
- Number of GCPs on each flight: 10, 7, and 5.

3.1. Data Collection

As we previously specified, the area chosen for the study is 178 m alongshore and 84 m cross-shore, and therefore its surface area is about 15,000 m^2, although the overflight area was obviously taken of a larger surface.

Data collection was planned with Phantom 4 Pro, based on the following three stages:

(a) Distribution on the beach of prefabricated landmarks to improve precision and calibration of the camera. Though some authors [27] state that direct georeferencing with high camera location accuracy and GNSS receivers can limit the need for GCPs, these landmarks were used as GCPs and georeferenced. The GCPs were plastic, measuring 24 × 30 cm and about 5 mm high (Figure 4), and had a hollow that helped fix their position in the sand. The reverse was painted, creating an alternating white-and-black grid. The positions were chosen to try to get an optimal placement according to the literature (covering all four corners of the site, the highest and lowest elevations, and with sufficient cross-shore and alongshore coverage). Moreover, their location was maintained during the two flight campaigns so that the results were not affected by any movement.

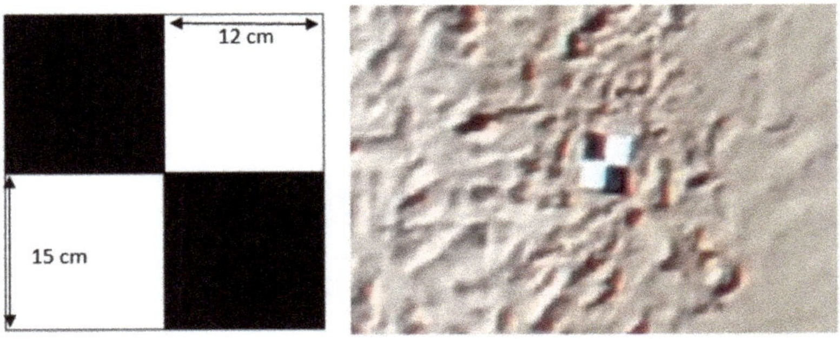

Figure 4. Dimensions of the GCP used and an aerial view of one of them already placed on the beach.

(b) Performing a topographic survey of the area using direct topography with GPS in RTK (Real-Time Kinematic) mode that provides precision around 3 cm. To ensure this accuracy, each topographic reading was repeated by taking three consecutive shots, which were checked and validated only if their difference was less than 1 cm in planimetry and 2 cm in altimetry. Moreover, the GPS rod man distinguished all the locations where pronounced changes in the beach topography appeared, due to his/her training and experience. Therefore, the density of GPS points was increased in these areas. A total of 657 survey points were taken, with an average distance among points of five meters approximately. This density of points is very high for the characteristics of the terrain.

The aims were twofold: first, calculating a topographic surface to compare with the photogrammetric data and, second, determining the coordinates of the GCPs for the RPAS postprocessing. The points were defined in European coordinates UTM ETRS89, and the levelings referred to the Spanish Datum (mean sea level in Alicante) by using the EGM2008 geoid provided by the National Geographic Institute [28].

(c) Introduction of the flight parameters into the RPAS software and realization of the photogrammetric flights. Six flights were made on the same day. In this way, the weather conditions and the situation of the terrain would be the same and therefore would not influence the results of the study. Flight planning requires the establishment of the limits of the area to be photographed, camera characteristics, flight height, flight direction, and frame overlap in side and forward directions. Thus, three flights were made at 8 am (when the sun angle is still low), with flight heights of 60, 80, and 100 m. The flights were taken with a side and forward overlap of 85%. However, also 75% of overlap in both directions was considered afterward by using the software. There were three more flights at noon (when the sun is at its highest position), repeating the same operation.

Therefore, six flights were made, but a total of 72 cases were studied (Table 1) by combining the parameters of two times of the day with different heights of flight (3), different side and forward overlaps and different number of GCPs (5, 7, and 10).

Table 1. Values of the different flight parameters and number of studied cases.

Parameter	Values	Number of Cases
Flight time	8 a.m. and 12 p.m.	2
Flight height	60, 80, and 100 m	3
Longitudinal overlap	70% and 85%	2
Transverse overlap	70% and 85%	2
Number of GCPs	5, 7, and 10	3
Total number of cases	-	72

Flight mission planning was previously done. For this, the Pix4D Capture program was used. This program uses the aerial images of Google Earth® as a base on which the area to be flown is defined, the flight height and the side and forward coverings were described, and the flight course was marked to optimize the times. The program calculated the flight speed and shooting interval among photographs. Figure 5 shows the flight plan scheme.

The camera technical data are presented in Table 2.

Table 2. Data of camera.

Sensor	1" CMOS
	Effective Pixels: 20M
Lens	FOV 84° 8.8 mm/24 mm (35 mm format equivalent) f/2.8−f/11 autofocus at 1 m-∞
Iso Range	Photo:
	100–3200 (Auto)
	8–1/2000 s
	8–1/8000 s
Mechanical Shutter Speed	3:2 Aspect Ratio 5472 × 3648
Electronic Shutter Speed	4:3 Aspect Ratio 4864 × 3648
Image Size	16:9 Aspect Ratio 5472 × 3078
	4096 × 2160 (4096 × 2160 24/25/30/48/50p)
	3840 × 2160 (3840 × 2160 24/25/30/48/50/60p)
PIV Image size	2720 × 1530 (2720 × 1530 24/25/30/48/50/60p)
	Single Shot
	Burst Shooting: 3/5/7/10/14 frames
Still Photography modes	Auto Exposure Bracketing (AEB): dL/5 at 0.7
	EV Bias Internal 2/3/5/7/10/15/20/30/60 s

Number of frames and duration of flight are shown in Table 3.

Table 3. Flight data, frame characteristics, frame number, and duration of flight.

Flight Height (m)	GSD (cm/pixel)	Frame Size (m·m)	Number of Frames	Duration of Flight (min)
60	1.64	90 × 60	185	16
80	2.18	119 × 80	123	11
100	2.73	148 × 98	81	9

The second column displays the Ground Sampling Distance (GSD), which is directly related to the flight altitude and camera parameters. The GSD is defined as the distance between two consecutive pixel centers measured on the ground. The greater the GSD value, the lower the spatial resolution of the image, and the less visible the details [29].

Figure 5. Image of the flight plan indicating the area of interest and frame overlap. The overflight area is bigger than the study area.

3.2. Method of Obtaining DTM by Photogrammetry and DTM Checking

The methodology for obtaining the DTM is based on the structure from motion (SfM) algorithm. The software used is Agisoft-Metashape Professional Educational®. Unstructured aerial images using fast, inexpensive, and highly automated image processing produces three-dimensional information. This RPAS-SfM pairing gives good results in cartographic production [27,30,31]

Firstly, once a set of frames was loaded into the software, an approximate orientation of the frames, based on the EXIF data of each photograph, was performed. EXIF is short for exchangeable image file, a format that is a standard for storing interchange information in digital photography image files using JPEG compression. It relied mainly on the focal length of the camera used, the time of taking the picture, and GPS coordinates.

Once the complete block was ordered, the program searched for tie points among adjacent frames. At this point, we could define the degree of precision that we require, as well as the key points and maximum tie points to be used in each frame to perform the operation.

The result of this process was a global point cloud that collected all the tie points of the flight frameset. At this time, the program had already created a three-dimensional point cloud. These point clouds were adjusted, georeferenced, and corrected for the lens distortion by using the GCPs. This procedure required entering the coordinates (X, Y, Z) of the GCPs and identifying them graphically in each of the frames in which they appeared. Since the GCPs points were defined in coordinates in the UTM-ETRS89 system, the adjusted point cloud would be in that same system.

At this point, we had the points of the topographic survey performed by GPS on the ground and the 72 DTMs obtained by processing the former set of point clouds. To facilitate and simplify the statistical reading of the DTMs, the size of each basic element of the DTM (tile size) was defined as a square of 1 m on the side. The Z value of each tile was defined as the average of the specific values it contained. Once the DTM was obtained, we cut it to the area of interest. By forming the DTM with all the points and cutting it later, we avoided the loss of data and extrapolation in the boundary areas.

These DTMs have been widely used, and much research on their error and uncertainty has already been investigated [14]. The quality of these models depends on several factors, such as the method used to attain the altimetric data, the density of the starting data, the resolution of the mesh, or the interpolation algorithm used, among others.

3.3. Calculation of the Error

To check the final quality of each flight, the error of each of the 72 DTMs (generated from the cloud of points obtained with the RPAS) was calculated by comparing to the DTM defined from the topographic data taken with GPS in RTK as the reference (Equation (1)). All the DTMs had the same dimensions, and a grid size of 1 × 1 m was chosen to facilitate comparison. Moreover, the percentage of grids that contained at least one datum was calculated, and its value was used as another reliability parameter.

The result of the comparison is another DTM whose characteristic is the difference between the altitudes of the flight DTM and the topographic DTM (GPS on the ground), that is, the vertical error (ε) in every grid.

$$\varepsilon = Z_{flight} - Z_{ground} \quad (1)$$

Figure 6 shows an example of this error (ε) calculated as the difference between the altitudes of the flight DTM and the topographic DTM (GPS on the ground)

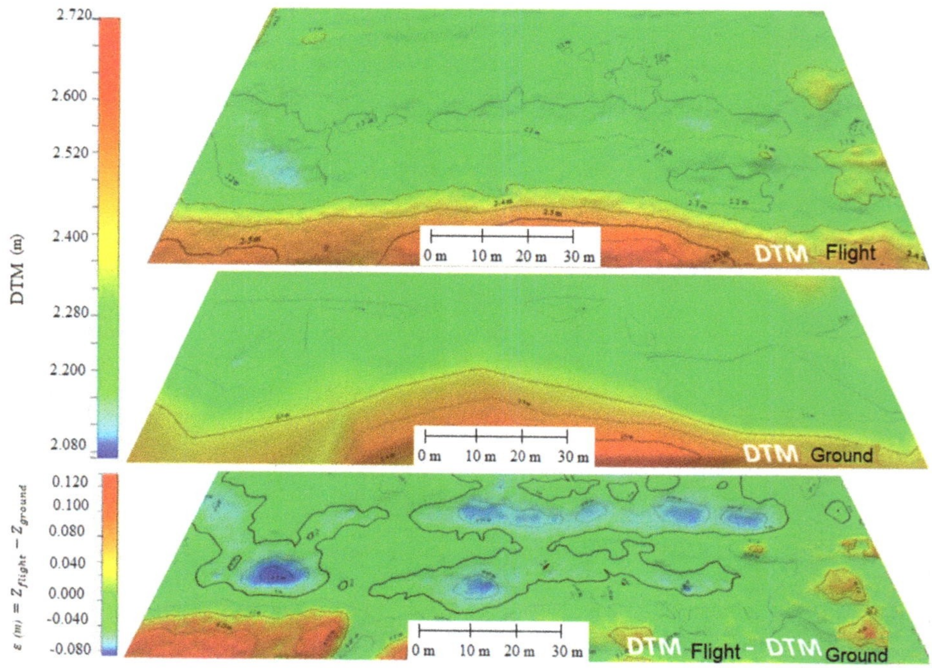

Figure 6. Map of the vertical error in every grid(ε). Example of the difference between the altitudes of the flight digital terrain models (DTM) and the topographic DTM (GPS on the ground).

Given the higher precision in the horizontal plane (approximately twice that in the vertical plane) and the very gentle slope of the beach profile (<2%), we will assume that the influence of the possible location error of a point on the vertical precision is negligible.

However, this average of the vertical errors suffers from that positives and negatives can cancel each other out and give a false sense of accuracy. That is the reason why another statistic, the RMSE (Equation (2)) was calculated.

The National Standard for Spatial Data Accuracy (NSSDA) is a recent standard proposed by the Federal Geographic Data Committee (1998) [32] and can be used for both analog and digital cartographic data [33]. This standard assumes a normal distribution of ε and uses the root-mean-square error (RMSE) as the most common and valid statistic for the evaluation of products obtained by photogrammetry and remote sensing.

$$RMSE_Z = \sqrt{\frac{1}{n}\sum_{i=1}^{n}\left(Z_{flight} - Z_{ground}\right)^2} \quad (2)$$

The 95% confidence interval (Equation (3)) for the vertical accuracy reached in each of the grids was determined according to the NSSDA as

$$P_{Z,95\%} = 1.96 \cdot RMSE_Z \quad (3)$$

Thus, Equation (4) shows the range of values that do not exceed the established accuracy.

$$\left\{ \begin{array}{c} \overline{x} + 1.96 \cdot RMSE_Z \\ \overline{x} - 1.96 \cdot RMSE_Z \end{array} \right\} \quad (4)$$

4. Results and Discussion

4.1. Error for Each of the 72 Cases

As previously mentioned, the average of the vertical errors (ε in Equation (1)) results in a number not too helpful because positives and negatives can cancel each other out. That is the reason why another statistic, the RMSE (Equation (2)) was calculated. From these data, the vertical accuracy (Equation (3)) for each of the cases was determined. The results of these two values for each of the 72 cases are shown in Table 4. Moreover, another error parameter defined in the methodology is also presented in Table 4, the percentage of non-empty grids (1×1 m^2), i.e., with at least one homogeneous point inside.

4.2. Influence of Number of GCPs

The first variable to consider is the number of GCPs. A box-and-whisker plot of their RMSE error is shown in Figure 7. Note that a boxplot is a standardized way of displaying the dataset based on a five-number summary: the minimum, the maximum, the sample median, and the first and third quartiles. You can see from the graph that there is a large distance between the lower (25%) and upper quartiles (75%) (IQR-Interquartile range), which are 0.10, 0.08, and 0.03 m for 5, 7, and 10 GCPs, respectively. Note that the whiskers (the two lines outside the box that extend to the highest and lowest observations) are similar in the three cases. The high value for the 10 GCP case is due to the existence of outliers for the noon survey. The variation of these results is far away from the results presented by other authors as James et al. [34] whose RMSE had a negligible deviation because of the number of GCPs, obtaining 3.12, 3.57, and 3.59 cm for 5, 10, and 15 GCPs, respectively.

Table 4. Different kinds of errors (RMSE, vertical accuracy, percentage of grids with homogeneous points) for data for different side (S) and forward (F) overlap and for different flight height, flight time, and number of GCPs.

Flight Time	Flight Height (m)	Type of Error	Overlap											
			85%F−85%S			85%F−70%S			70%F−85%S			70%F−70%S		
			Number of GCPs											
			10	7 GCP	5 GCP	10 GCP	7 GCP	5 GCP	10 GCP	7 GCP	5 GCP	10 GCP	7 GCP	5 GCP
8 a.m.	60	RMSE (m)	0.03	0.05	0.05	0.04	0.04	0.05	0.03	0.06	0.05	0.03	0.03	0.05
		Vertical accuracy (1.96·RMSE)	0.06	0.10	0.10	0.08	0.08	0.10	0.06	0.12	0.10	0.06	0.06	0.10
		% grid with data		65.80%			59.73%			59.73%			60.29%	
	80	RMSE (m)	0.04	0.04	0.04	0.04	0.04	0.04	0.04	0.04	0.04	0.04	0.04	0.05
		Vertical accuracy (1.96·RMSE)	0.08	0.08	0.08	0.08	0.08	0.08	0.08	0.08	0.08	0.08	0.08	0.10
		% grid with data		60.45%			60.15%			60.85%			55.83%	
	100	RMSE (m)	0.05	0.07	0.09	0.05	0.05	0.09	0.05	0.05	0.08	0.05	0.05	0.19
		Vertical accuracy (1.96·RMSE)	0.10	0.14	0.18	0.10	0.10	0.18	0.10	0.10	0.16	0.10	0.10	0.37
		% grid with data		46.00%			36.00%			36.00%			35.97%	
12 p.m.	60	RMSE (m)	0.06	0.05	0.05	0.10	0.14	0.15	0.033	0.04	0.04	0.14	0.15	0.16
		Vertical accuracy (1.96·RMSE)	0.12	0.10	0.10	0.20	0.27	0.29	0.06	0.08	0.08	0.27	0.29	0.31
		% grid with data		43.24%			39.26%			37.60%			38.53%	
	80	RMSE (m)	0.05	0.05	0.05	0.10	0.14	0.16	0.05	0.05	0.05	0.145	0.17	0.17
		Vertical accuracy (1.96·RMSE)	0.10	0.10	0.10	0.20	0.27	0.31	0.10	0.10	0.10	0.27	0.33	0.33
		% grid with data		39.03%			36.17%			36.95%			33.47%	
	100	RMSE (m)	0.06	0.12	0.13	0.07	0.04	0.09	0.06	0.07	0.07	0.15	0.18	0.19
		Vertical accuracy (1.96·RMSE)	0.12	0.24	0.25	0.14	0.08	0.18	0.12	0.14	0.14	0.29	0.355	0.37
		% grid with data		29.66%			20.47%			21.37%			19.18%	

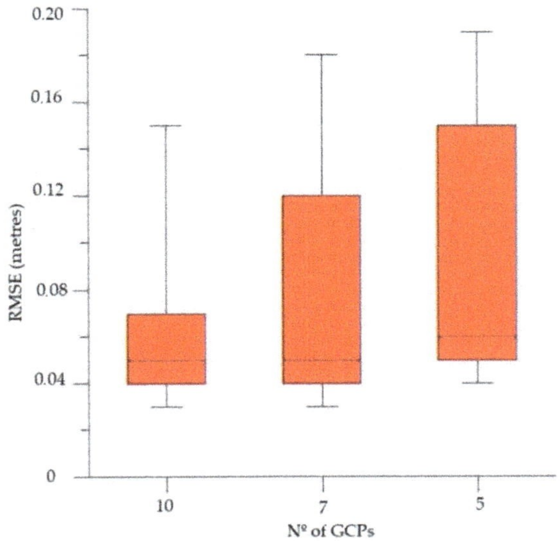

Figure 7. Box-and-whisker plot of the average error based on the GCP number.

Thus, though other authors such as Zimmerman et al. [30] found that (7 to 9) well-placed GCPs in the optimal configuration produced the same magnitude of error as using more (15) poorly placed GCPs, the only acceptable values in our case are those collected by using 10 GCPs, with an IQR less than 3 cm.

Moreover, 10 GCPs is just the maximum number of points for this particular case. To generalize this value and its use in any other case, the number of GCPs has been divided by the surface in hectares (Ha), a unit frequently used in topographic surveys. Thus, if we divide 10 GCPs by 1.5 Ha (15,000 m^2), we get a rounded value for the density of GCPs (7 GCP/Ha), a new starting parameter when designing a beach leveling campaign using RPAS. Regrettably, a limitation of this study is that we did not check whether the accuracy might even increase more by using more than 10 GCP, and, therefore, the trend of the inclusion of more GCPs remains unknown.

4.3. *Influence of Flight Time*

As previously noted, visual homogeneity of beach areas is one of the main problems for surveying using photogrammetric methods because of the reduction of homologous points among adjacent frames. Therefore, two different times for the flights were chosen (8 a.m. and 12 a.m.) to find out if shadows in the early morning (Figure 8) produce more homologous points than at noon and, thus, a decrease in the error committed. The number of tie points are presented in Table 5.

Based on the results of the previous subsection, the values obtained for 5 and 7 GCPs were discarded to avoid distorting the statistical results of the rest of the variables. Figure 7 shows the values of vertical RMSE for both aforementioned times of the flight and all the flight heights

As can be seen in Figure 8 the IQR ranges from 5 to 14 cm with a median of 0.09 m at noon while there are just a mean of 3.5 cm, no outliers, and a negligible dispersion in the early morning. Moreover, the percentage of grids with data is almost 60% higher for early flights than for noon flights. Therefore, it can be established that RPAS for beach leveling must be performed early in the morning. Or, in other words, RPAS surveys must be banned at noon.

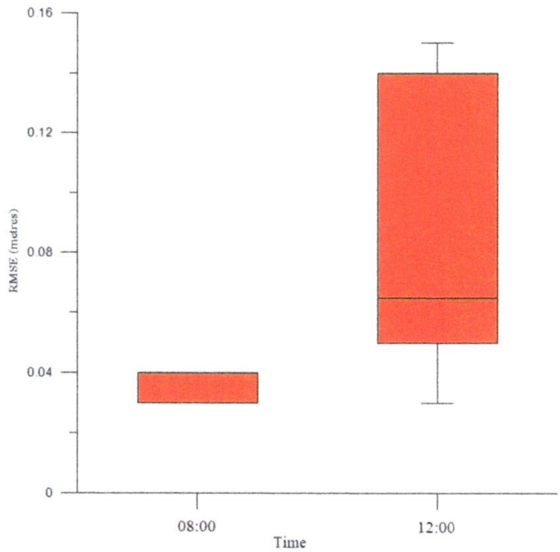

Figure 8. Box-and-whisker plot of RMSE vs. the flight time (8 a.m. and 12 a.m.) for all the flight heights.

Table 5. Number of tie points.

Flight Time	Flight Height	Number of Tie Points
	60 m	105,382
08:00 a.m.	80 m	76,852
	100 m	32,733
	60 m	39,296
12:00 a.m.	80 m	29,639
	100 m	19,848

4.4. Influence of Frame Overlap and Flight Height

Analyzing the percentage (70% vs. 85%) of side and forward overlap, four different cases were considered. Moreover, flights were carried out at three different heights (60, 80, and 100 m). The results presented in Table 4 are now shown in Figure 9, where only the 10 GCP experiments have been taken into account.

Starting with the noon flights (dashed lines), it can be seen that there is not too much difference between the results of 60 and 80 m flight heights. Regarding the forward overlap, there is no appreciable difference between both (70 and 85) percentages, i.e., forward overlap change from 85 to 70 did not influence final results. However, RMSE decreased enormously (from 0.15 to 0.05 m) when the side or transverse overlap changed from 70% to 85%.

On the other side, when results from 8 am flights are analyzed, RMSE remained constant and, therefore, independent from both side and forward overlap percentages. Furthermore, there is a small but still significant difference for RMSE as a function of the flight height. RMSE was 3, 4, and 5 cm for 60, 80, and 100 m heights, respectively. Following Gonçalves and Henriques [35], a relative accuracy of the flying height can be calculated. This relative accuracy was 0.35‰ in their case (from 0.046 m to 131 m flying height). A numerical value very similar to the results presented here where relative accuracy was 0.5‰ for each of the flights performed early in the morning but lesser than the values found for the flights performed at noon which can reach up to 1.5‰.

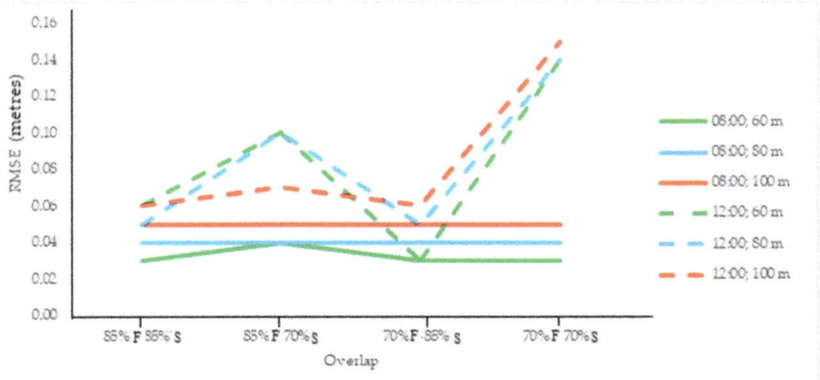

Figure 9. RMSE vs. different side (S) and forward (F) overlap for different flight heights and flight times (dashed lines are used for noon flights while solid lines are for 8 a.m. flights). Note that the number of GCPs is not a variable because only experiments performed with 10 GCPs were considered.

Thus, in brief, it can be established again that early morning flights minimize vertical error. Moreover, side overlap should not be less than 85% while forward overlap percentage is not a decisive factor. Finally, the decision about the flight height (when designing an RPAS for a beach leveling) must take into account that variation of vertical RMSE, though small in absolute value (5 cm vs. 3 cm), can be relatively significant (about 66%) when a 100 m height is chosen instead of a 60 m height.

5. Conclusions

A common fact in the photogrammetry of beaches (poorly differentiated visual areas) is the difficulty in the identification of common points in contiguous frames [22]. And, obviously, the fewer the homologous points defined by the program, the lower the accuracy. Thus, the main objective of this work is to determine the parameters of flight (height, time, frame overlap) and number of GCPs to optimize the accuracy of photogrammetric surveys when using RPAS in cases of visually homogeneous areas.

The following variables have been taken into account: flight height (60, 80, and 100 m), flight time (8 a.m. and 12 p.m.), side and forward overlap (70% vs. 85%), and the number of ground control points or GCPs. The combination of these variables results in 72 cases.

Firstly, one of the main conclusions is related to the density of GCPs. A minimum value of 7 GCPs/Ha has been found for this new parameter when designing a beach leveling campaign using RPAS. However, the trend of the inclusion of more GCPs remains unknown. This aspect is pending for future research.

Secondly, there is no appreciable difference regarding the forward overlap. But, on the other side, RMSE increased to three times (from 0.05 to 0.15 m) when side overlap decreased from 85% to 70%.

Moreover, the median of the error for noon flights (7 cm) is double that for the early morning flights (3.5 cm) because of the higher (almost 60%) percentage of grids with data for early flights. Therefore, beach levelings must never be performed at noon when carried out by RPAS.

Finally, there is a significant difference (till 66%) for RMSE as a function of the flight height. RMSE was 3, 4, and 5 cm for 60, 80, and 100 m heights, respectively, when only results from the 8 a.m. flights are analyzed. Furthermore, in this case, RMSE remains constant, and therefore independent, for the different side and forward overlap percentages.

Author Contributions: Conceptualisation, F.C.-d.-V.; F.J.G. and J.J.M.-P.; investigation and writing—original draft preparation, F.C.-d.-V.; F.J.G.; J.J.M.-P.; A.C.-d.-V.; V.R.-O.; P.L.; S.G.-L.; B.J.; review and editing, F.C.-d.-V.; J.J.M.-P. All authors have read and agreed to the published version of the manuscript.

Funding: This research was partially funded by Fundacion Campus Tecnologico de Algeciras. And the APC was funded by the Coastal Engineering Research group (University of Cadiz).

Informed Consent Statement: Not applicable.

Data Availability Statement: Data sharing not applicable.

Conflicts of Interest: The authors declare no conflict of interest.

References

1. Muñoz-Perez, J.J.; Medina, R. Comparison of long-, medium- and short-term variations of beach profiles with and without submerged geological control. *Coast. Eng.* **2010**, *57*, 241–251. [CrossRef]
2. Houston, J. International tourism and US beaches. *ShoreBeach. Shore Beach* **1996**, *64*, 3–4.
3. Jigena, B.; de Gil, A.; Walliser, J.; Vidal, J.; Muñoz-Perez, J.J.; Pozo, L.; Lebrato, J. Improving the Learning Process in the Subject of Basic Maritime Training Using Gps and Google Earth As Useful Tools. *INTED Proc.* **2016**, *1*, 6161–6171. [CrossRef]
4. Berrocoso, M.; Páez, R.; Jigena, B.; Cartula, C. The RAP Net: A Geodetic Positioning Network for Andalusia (South Spain). In Proceedings of the EUREF Publication No. 16, Mitteilungen des Bundesamtes für Kartographie und Geodäsie, Riga, Latvia, 14–16 June 2006; pp. 364–368.
5. Payo, A.; Favis-Mortlock, D.; Dickson, M.; Hall, J.W.; Hurst, M.D.; Walkden, M.J.; Townend, I.; Ives, M.C.; Nicholls, R.J.; Ellis, M.A. Coastal Modelling Environment version 1.0: A framework for integrating landform-specific component models in order to simulate decadal to centennial morphological changes on complex coasts. *Geosci. Model Dev.* **2017**, *10*, 2715–2740. [CrossRef]
6. Burdziakowski, P.; Specht, C.; Dabrowski, P.S.; Specht, M.; Lewicka, O.; Makar, A. Using UAV photogrammetry to analyse changes in the coastal zone based on the sopot tombolo (Salient) measurement project. *Sensors (Switzerland)* **2020**, *20*, 4000. [CrossRef] [PubMed]
7. Contreras de Villar, A.; Gómez-Pina, G.; Muñoz-Pérez, J.J.; Contreras, F.; López-García, P.; Ruiz-Ortiz, V. New design parameters for biparabolic beach profiles (SW Cadiz, Spain). *Rev. Constr.* **2019**, *18*, 432–444. [CrossRef]
8. Specht, C.; Lewicka, O.; Specht, M.; Dabrowski, P.; Burdziakowski, P. Methodology for carrying out measurements of the tombolo geomorphic landform using unmanned aerial and surface vehicles near Sopot Pier, Poland. *J. Mar. Sci. Eng.* **2020**, *8*, 384. [CrossRef]
9. Valderrama, L.; Dubois, R.M.; Ressl, R.; Silva, R.; Cruz, C.; Muñoz-Pérez, J. Dynamics of coastline changes in Mexico. *J. Geogr. Sci.* **2019**, 1–18. [CrossRef]
10. Muñoz-Perez, J.J.; Payo, A.; Roman-Sierra, J.; Navarro, M.; Moreno, L. Optimization of beach profile spacing: An applicable tool for coastal monitoring. *Sci. Mar.* **2012**, *76*, 791–798. [CrossRef]
11. Castelle, B.; Laporte-Fauret, Q.; Marieu, V.; Michalet, R.; Rosebery, D.; Bujan, S.; Lubac, B.; Bernard, J.B.; Valance, A.; Dupont, P.; et al. Nature-based solution along high-energy eroding sandy coasts: Preliminary tests on the reinstatement of natural dynamics in reprofiled coastal dunes. *Water (Switzerland)* **2019**, *11*, 2518. [CrossRef]
12. Payo, A.; Kobayashi, N.; Muñoz-Pérez, J.; Yamada, F. Scarping predictability of sandy beaches in a multidirectional wave basin. *Ciencias Mar.* **2008**, *34*, 45–54. [CrossRef]
13. Laporte-Fauret, Q.; Marieu, V.; Castelle, B.; Michalet, R.; Bujan, S.; Rosebery, D. Low-Cost UAV for high-resolution and large-scale coastal dune change monitoring using photogrammetry. *J. Mar. Sci. Eng.* **2019**, *7*, 63. [CrossRef]
14. Ojeda, J.; Vallejo, I.; Malvarez, G.C. Morphometric evolution of the active dunes system of the Doñana National Park, Southern Spain (1977–1999). *J. Coast. Res.* **2005**, *49*, 40–45.
15. Talavera, L.; del Río, L.; Benavente, J.; Barbero, L.; López-Ramírez, J.A. UAS & S f M-based approach to Monitor Overwash Dynamics and Beach Evolution in a Sandy Spit. *J. Coast. Res.* **2018**, *85*, 221–225. [CrossRef]
16. Shervais, K.A.H.; Kirkpatrick, J.D. Smoothing and re-roughening processes: The geometric evolution of a single fault zone. *J. Struct. Geol.* **2016**, *91*, 130–143. [CrossRef]
17. James, M.R.; Robson, S.; Smith, M.W. 3-D uncertainty-based topographic change detection with structure-from-motion photogrammetry: Precision maps for ground control and directly georeferenced surveys. *Earth Surf. Process. Landf.* **2017**, *42*, 1769–1788. [CrossRef]
18. Gabara, G.; Sawicki, P. Multi-variant accuracy evaluation of UAV imaging surveys: A case study on investment area. *Sensors (Switzerland)* **2019**, *19*, 5229. [CrossRef]
19. Fisher, P.F.; Tate, N.J. Causes and consequences of error in digital elevation models. *Prog. Phys. Geogr.* **2006**, *30*, 467–489. [CrossRef]
20. Hugenholtz, C.H.; Whitehead, K.; Brown, O.W.; Barchyn, T.E.; Moorman, B.J.; LeClair, A.; Riddell, K.; Hamilton, T. Geomorphological mapping with a small unmanned aircraft system (sUAS): Feature detection and accuracy assessment of a photogrammetrically-derived digital terrain model. *Geomorphology* **2013**, *194*, 16–24. [CrossRef]

21. Gindraux, S.; Boesch, R.; Farinotti, D. Accuracy assessment of digital surface models from Unmanned Aerial Vehicles' imagery on glaciers. *Remote Sens.* **2017**, *9*, 186. [CrossRef]
22. Barbero, J.J.; García-López, L.; López-Ramírez, S.; Muñoz, J.A. RPAS as a New Tool for the Study of Sand Dunes in Coastal Environments: A Case Study in the South Atlantic Area of Spain; Vila real (Portugal) 2017. Available online: http://uas4enviro2017.utad.pt/wp-content/uploads/2017/10/Abstract_Book_UAS4Enviro2017_completo.pdf (accessed on 30 October 2020).
23. Li, Z.; Zhu, Q.; Gold, C. *Digital Terrain Modeling Principles and Methodology*, 1st ed.; CRC Press: Boca Raton, FL, USA, 2004; ISBN 9780415324625.
24. Puertos del Estado. Available online: http://www.puertos.es/es-es/oceanografia/Paginas/portus.aspx (accessed on 30 October 2020).
25. Hidtma, S.L.; UTE Ecoatlántico. Estudio Ecocartográfico del Litoral de la Provincia de Cádiz. Ministry of the Environment Ref. 28-4983. 2013. Available online: https://www.miteco.gob.es/es/costas/temas/proteccion-costa/ecocartografias/ecocartografia-cadiz.aspx (accessed on 30 October 2020).
26. Moloney, J.G.; Hilton, M.J.; Sirguey, P.; Simons-Smith, T. Coastal Dune Surveying Using a Low-Cost Remotely Piloted Aerial System (RPAS). *J. Coast. Res.* **2018**, *34*, 1244–1255. [CrossRef]
27. Taddia, Y.; Stecchi, F.; Pellegrinelli, A. Coastal Mapping Using DJI Phantom 4 RTK in Post-Processing Kinematic Mode. *Drones* **2020**, *4*, 9. [CrossRef]
28. de Fomento, M. Instituto Geográfico Nacional. Available online: https://www.ign.es/web/ign/portal (accessed on 30 October 2020).
29. Ground Sampling Distance (GSD). Available online: https://support.pix4d.com/hc/en-us/articles/202559809-Ground-sampling-distance-GSD (accessed on 29 January 2020).
30. Zimmerman, T.; Jansen, K.; Miller, J. Analysis of UAS Flight Altitude and Ground Control Point Parameters on DEM Accuracy along a Complex, Developed Coastline. *Remote Sens.* **2020**, *12*, 2305. [CrossRef]
31. James, M.R.; Chandler, J.H.; Eltner, A.; Fraser, C.; Miller, P.E.; Mills, J.P.; Noble, T.; Robson, S.; Lane, S.N. Guidelines on the use of structure-from-motion photogrammetry in geomorphic research. *Earth Surf. Process. Landforms* **2019**, *44*, 2081–2084. [CrossRef]
32. National Standard for Spatial Data Accuracy. 1998. Available online: https://www.fgdc.gov/standards/projects/accuracy/part3/chapter3 (accessed on 30 October 2020).
33. Ruiz-Lendínez, J.J.; Ariza-López, F.J.; Ureña-Cámara, M.A. Study of NSSDA variability by means of automatic positional accuracy assessment methods. *ISPRS Int. J. Geo-Inf.* **2019**, *8*, 552. [CrossRef]
34. James, M.R.; Robson, S.; d'Oleire-Oltmanns, S.; Niethammer, U. Optimising UAV topographic surveys processed with structure-from-motion: Ground control quality, quantity and bundle adjustment. *Geomorphology* **2017**, *280*, 51–66. [CrossRef]
35. Gonçalves, J.A.; Henriques, R. UAV photogrammetry for topographic monitoring of coastal areas. *ISPRS J. Photogramm. Remote Sens.* **2015**, *104*, 101–111. [CrossRef]

Journal of
Marine Science and Engineering

MDPI

Technical Note

A New Method for the Collection of Marine Geomagnetic Information: Survey Application in the Colombian Caribbean

Karem Oviedo Prada [1,2,*], Bismarck Jigena Antelo [1,*], Nathalia Otálora Murillo [2], Jeanette Romero Cózar [1], Francisco Contreras-de-Villar [1] and Juan José Muñoz-Pérez [1,*]

[1] Puerto Real Campus, University of Cadiz, 11510 Puerto Real (Cadiz), Spain; jeanette.romero@uca.es (J.R.C.); francisco.contreras@uca.es (F.C.-d.-V.)

[2] Oceanographic and Hydrographic Research Centre of the Caribbean, Barrio Bosque, Sector Manzanillo, Escuela Naval de Cadetes "Almirante Padilla", Cartagena de Indias 130001, Colombia; notalora@dimar.mil.co

* Correspondence: ing.karemoviedo@gmail.com (K.O.P.); bismarck.jigena@gm.uca.es (B.J.A.); juanjose.munoz@uca.es (J.J.M.-P.)

Abstract: In recent years, the Oceanographic and Hydrographic Research Center (part of the General Maritime Directorate of Colombia (DIMAR) has made important efforts to advance research in the field of marine geophysics, in particular, the techniques of geomagnetism, sub-bottom profiling, and side-scan sonar, the first being the most developed at the present time. A method is presented for the acquisition of geomagnetic data in marine environments, as used by DIMAR in the Colombian maritime territory. The development of the geomagnetic method not only offers the opportunity to advance basic scientific knowledge, but it is also of great importance in support of national sovereignty issues. Among other applications, the most representative uses of the geomagnetic method are the location of pipelines and metal plates, detection of buried ordnance, identification of sites of archaeological interest, and the identification and characterization of geological structures. As a result of testing the method, a grid of geomagnetic data was surveyed in an area close to the Island of San Andrés in the north-west of the Colombian maritime territory. The survey was prepared with a regional geometric arrangement, the result of which was compared with survey data obtained from the National Oceanic and Atmospheric Administration (NOAA) magnetic data repository and carried out in the same study area. Despite the long time interval between the two surveys, almost 50 years, no significant differences were observed in terms of the analyzed variables. Finally, results show negligible differences between the magnetic data obtained for the years 1970 and 2018 for all the variables measured, such as the inclination, declination, and total magnetic field. These differences may be attributable to a geological component or also to the acquisition and processing methods used in the 1970s.

Keywords: marine geophysics; magnetic method; Colombian Caribbean; DIMAR; CIOH

check for updates

Citation: Oviedo Prada, K.; Jigena Antelo, B.; Otálora Murillo, N.; Romero Cózar, J.; Contreras-de-Villar, F.; Muñoz-Pérez, J.J. A New Method for the Collection of Marine Geomagnetic Information: Survey Application in the Colombian Caribbean. *J. Mar. Sci. Eng.* **2021**, *9*, 10. https://dx.doi.org/10.3390/jmse9010010

Received: 20 November 2020
Accepted: 16 December 2020
Published: 23 December 2020

1. Introduction

The increase in marine geophysical activity in recent years has provided essential data for evaluating theories about the origin of oceans and continents. Of the different methods used to explore the sea floor and underlying mantle, the magnetic field and its measurements have proven to be one of the most powerful tools for discovering and delineating structural and geological patterns [1].

According to Ewing et al. [2], before World War II, almost all marine magnetic observations had been made by the research ship "Carnegie" (1909–1929), which was specially built to work along widely spaced lines in the Atlantic, Pacific and Indian oceans. After the war, the fluxgate magnetometer, originally developed as an airborne instrument for detecting submarines, was adapted for marine applications by the Lamont Geological Research Observatory. These were the first measurements made with a magnetometer towed by a ship. Later, for work in the maritime field, the fluxgate magnetometer was replaced by the proton magnetometer, having the advantages of absolute field measurement and not requiring orientation of the head [3].

The use of the geomagnetic method is widely known globally, for its various local and regional applications [4]. Due to its high costs, which involve equipment and logistic development and multiple applications, this geophysical method is generally undertaken by foreign private companies with commercial aims, as the high costs prove to be a disadvantage for state organizations dedicated to science and marine research, which have limited budgetary allocations [5]. Therefore, it can be summarized that established foreign companies, which compose the entire market, dominate geophysical exploration, including those services in limited supply, such as gravimetry and magnetometry and its applications [6–8].

The General Maritime Directorate (DIMAR) is located at Cartagena de Indias, in the Colombian Caribbean Sea. DIMAR started the project "Geomagnetismo Marino" in 2015 with the purpose of recovering research capacity through the use of the G-882 marine magnetometer from geometrics. One of the recovery activities included training on the handling of the magnetic sensor and data acquisition. For the former, a document was produced [9] in which a vast database and manuals were compiled, which served as a base for the production of the following geophysical work methodology.

The need to propose a work methodology was pressing, as there was no record in Colombia of any other public entity carrying out this type of scientific research. Therefore, the efforts of the Caribbean Oceanographic and Hydrographic Research Centre (CIOH) were aimed at the standardization of the guidelines and the parameters required for the optimization of the marine geomagnetic method. After much effort and field tests in the Colombian Caribbean, a methodology has been obtained that offers high-quality marine geomagnetic data collection.

Thus, the following method aims to sequentially show the planning and acquisition of geomagnetic information in deep marine environments in Colombian territory on board the oceanographic research vessel ARC (Navy of the Republic of Colombia) Providence. The guide has become a tool that provides an effective and efficient response for geophysical research at the service of the nation. For this work, a bibliographic compilation was carried out taking into account aspects such as the verification of magnetic sensors and operators that can be powerful sources of magnetic noise. [10]. A fundamental aspect in the survey was to determine the distance at which the magnetometer sensor must be towed to reduce the magnetic effects of the vessel. Finally, the optimal lateral spacing between the lines also had to be considered, which is directly related to the depth of the water [11].

Moreover, geomagnetism is a geophysical prospecting method, applicable to the oil industry, and also mining and archaeological artefact explorations [11–13]. In mineral exploration, magnetometry is widely used to directly prospect for magnetic minerals, such as magnetite and other ferromagnetic minerals, and the method stands out for its speed and low cost. This method is the most widely used in geophysical surveys, at local and regional scales, and it is based on the study of the Earth's magnetic field and its variations, as a consequence of additional magnetic fields produced by magnetized rocky bodies positioned on the surface and close subsoil [14,15].

The magnitude measured in the magnetic method is the Geomagnetic Field, which is related to the magnetization of the environment and which, in the majority of materials, appears when a magnetic field is applied to a body [16]. In the magnetic method, the objective is to investigate the geology of the subsoil, from the variations in this geomagnetic field, resulting from the magnetic properties of the underlying rocks [17,18]. Not all the rock-forming minerals are magnetic, but certain types of rock contain sufficient magnetic minerals to be able to produce significant magnetic anomalies, such as iron and magnetite, among others. The influence of the total magnetic field can be measured anywhere on earth, with a certain direction and intensity, subject to periodic variations and non-periodic disturbances, the magnitude of which on the planet's surface can vary from point to point from 25,000 to 65,000 nT [19,20].

When a magnetic material is placed in a magnetic field, the material is magnetized, and the external field of magnetization is reinforced with the induced magnetic field in the material. This is known as induced magnetization, and it is based on the magnetic susceptibility of the materials, (understood as the degree of magnetization of a material in response to a magnetic field), and the magnitude and direction of the magnetic field [21]. When the external field disappears, the induced magnetization disappears immediately, but some materials retain a residual magnetism, and its direction will be fixed in the direction of the inductive field [22]. The residual magnetism reflects the history of the material. Thus, there is a contrast of magnetism between an anomalous source and the adjacent lateral formations. These two types of magnetization are due to spontaneous magnetization, which is a property of the ferromagnetic minerals in the Earth's crust [23].

To calibrate the data, check its reliability, and study the variation of the new data, it was necessary to have a reference work. To do this, we took into account previous work carried out between 1970 and 1971, obtained from the repository of the National Oceanic and Atmospheric Administration (NOAA). More specifically, from the Marine Geology and Geophysics data from the National Centers for Environmental Information (NCEI), formerly the National Geophysical Data Center or NGDC [24].

The objective of this work is to present a method to carry out magnetic surveys that is compatible with other techniques used in different areas of engineering and science (hydrographic surveys, side-scan sonar, search for magnetized bodies, search for archaeological remains, etc.). To evaluate the quality of the work, the results obtained in a recent campaign were compared with those from previous ones, and the differences were analyzed.

2. Materials and Methods

2.1. Study Area

The study area is located in the Colombian Caribbean Sea and more specifically to the south of the archipelago of San Andrés, Providencia and Santa Catalina (SAPSC). The geophysical survey was carried out in the area located between San Andrés Island, Cayos de Albuquerque Island and Cayos de Este-Sudeste Island, within the polygon marked in yellow, as can be seen in Figure 1. The study area covers an area of approximately 2040 km^2.

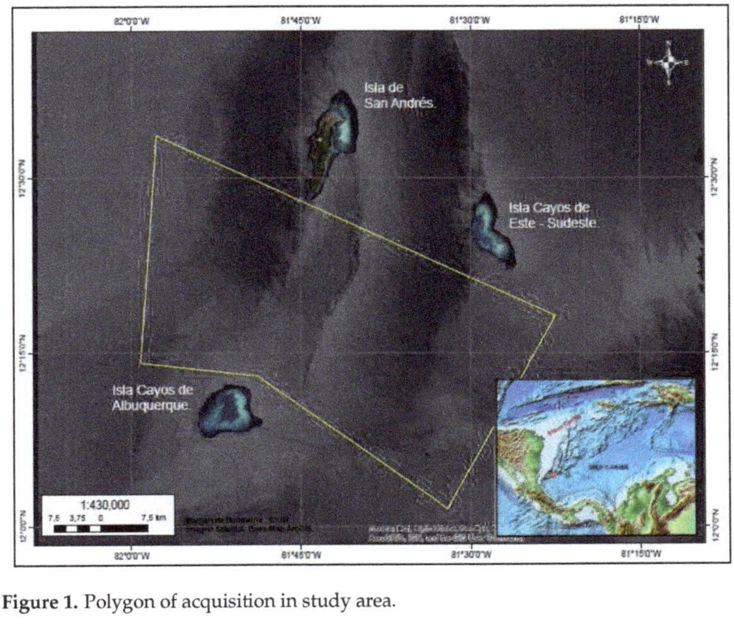

Figure 1. Polygon of acquisition in study area.

A magnetic survey measures the local magnetic field characteristics of a certain region. This type of technology only detects minerals and/or materials that respond to magnetic fields. For this reason, its applications are mainly aimed at mineral exploration, but it can also be useful for the exploration of coal, oil, and gas and in the detection of shipwrecks. [2,6,25,26]. A geophysical survey consists of different phases.

2.2. PHASE 1. Planning of the Acquisition Campaign

The form of the geophysical survey is established in this phase, and the times, the necessary inputs, and the possible unforeseen events that may occur at sea, are estimated.

Before planning the data acquisition, the study objective and the scale of the work (local or regional) should initially be taken into account. The configuration and length of the lines to be acquired will depend on these.

The generation of the acquisition grid is made based on the sought objectives. It is important to consider whether it is required to determine the regional magnetic field (e.g., changes of magnetic polarity reflected in the marine magnetic anomalies, regional guidelines, etc.), or to determine local geologic anomalies (e.g., geologic bodies and structures), or to identify the anomalies due to metallic objects produced by humans [27]. This is related to the fact that the geometric arrangement of the acquisition must take into account the spatial resolution of the body to be characterized—that is, the smaller the object, the denser and less spaced the survey grid must be.

It is also important to take into account the sensitivity of the sensors, since, to recognize an anomaly, this must be several times greater than the sensitivity (resolution) of the magnetometer and the external noise level. It is important to define this parameter to know if the object is detectable on the surface and, in such a case, how much the readings in the profile, and the distance between adjacent profiles, would have to be spaced (spacing of the grid). Ideally, a grid should be shaped to cover the whole area in such a way that the anomaly can be always detected by a profile. This means that there must be some overlapping between profiles [9].

Additional magnetic information is required, whether from magnetic observatories or from a Base Station near the survey area, with the purpose of improving the quality of the data. In this case, a Geometrics G-862 RBS Base Station (Figure 2) was used, which was acquired by the General Maritime Directorate in 2015. This was positioned at a minimum radius of 60 m from any source of electromagnetic interference. This reduces the errors that can occur in the data, due to fortuitous cases, such as electromagnetic interference from the solar field.

Figure 2. Installation of the Geometrics G-862 RBS Base Station.

Starting the planning activity, it is essential to have high-resolution bathymetric data, in order to support the identification of the geological structure that is required to be recorded with magnetometry [28]. As the objective of the project was to determine the magnetic anomalies, generated by the volcanic bodies and geological structures (faults) located to the south of the SAPSC, in the vicinity of San Andrés Island, Cayos de Albuquerque Island and Cayos de Este- Sudeste Island (Figure 1). The survey lines were carried out, taking into account the geoforms displayed in the bathymetry. For this reason, the lines are established perpendicular to faults or other structures, in regular meshes, where it is ensured that the separation between lines is equal to the estimated minimum distance between the sensor and the magnetic object or target. This is why it is recommended to take into account the depths at which the survey will be carried out [29].

The area included in the geophysical research polygon, in which it was planned to undertake five main lines of acquisition, with a NW-SE direction (azimuth of 300°), with lengths between 70 and 57 km, and a separation of 7 km. The six control lines, oriented perpendicularly to the main lines, are distributed with a spacing of 23.50 km, and they have of an average length of 32 km (Figure 3). In order to calculate the days needed for the survey, the total length of the lines at the optimal survey speed in linear nautical miles was considered, assuming 24 working hours per day [9]. This calculation is shown in Tables 1 and 2.

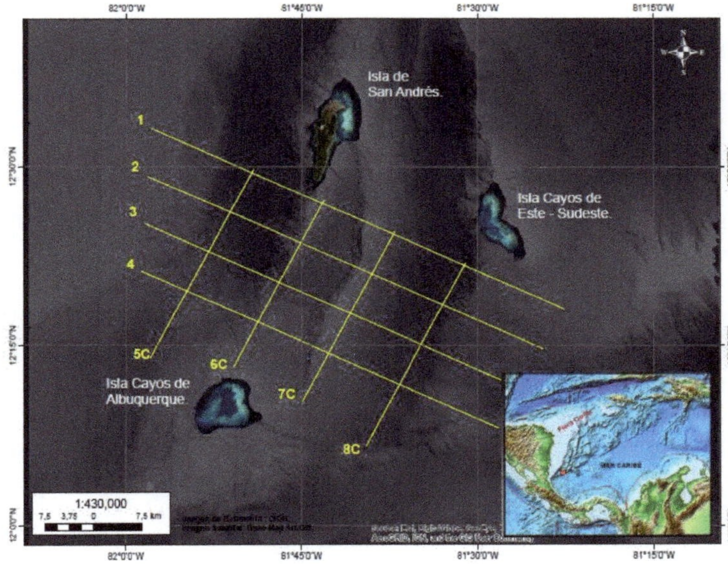

Figure 3. Lines of survey.

Table 1. Estimation of duration of the survey in days from linear nautical miles.

Survey	Number of Lines	Meters	LNM	Time in Days
San Andrés	4	317,087	171.213	2
Control lines	4	92,784	92.784	1
SURVEY				
120 LNM = 24 h 263.99 LNM = 3 Days of survey				

Once the configuration of the survey was established, the times for the voyages and duration of the acquisition were estimated. As mentioned previously, it is important to maintain good data density that can adequately represent the objective; that is to say, the optimum survey speed was 5 knots, assuring that the intensity of the signal was stable, and 10 samples per second were obtained.

Finally, a meticulous control was made to ensure that the acquisition of the data was carried out successfully. In the case of consumable equipment, such as RS232-USB converters, it is ideal to have spare parts, in case of unexpected events.

Table 2. Estimation of time of the operation, including the displacement.

Activity	Days	Start Date	End Date
Voyage: Cartagena to San Andrés Island	02	D	D+2
Execution of the magnetometry survey in the deep waters off San Andrés Island	03	D+3	D+6
Voyage: Study area to Cartagena	02	D+2	D+8
		TOTAL DAYS OF OPERATION:	08

2.3. PHASE 2. Data Acquisition

The oceanographic research vessels, ARC Malpelo and ARC Providencia, were enabled to operate with the Geometrics G-882 marine magnetometer [30], property of the DIMAR (Figure 4). This apparatus has a broad range of detection for ferrous materials of various sizes and a sensitivity of <0.004 $nT/\pi Hz$ rms, which increases the probability of detection. It has a hydrodynamic design that helps reduce the probability of rock incrustation, and it operates to a depth of approximately 2750 m, and at temperatures from −35 °C to 50 °C. The cesium–vapor sensor is at the rear of the "fish" in the cylinder that forms a T with the longest axis, where the direction of the sensor can be modified; this was vertical, as the work was to be carried out in equatorial latitudes. Finally, the sampling interval ranged from one sample every three seconds, to twenty samples per second, with an absolute precision of <2 nT. The acquisition of field data was carried out with MagLog software from Geometrics Inc.

Figure 4. Geometrics G-882 marine magnetometer. Source Karem Oviedo Prada, 2020.

2.4. PHASE 3. Office and Data Processing

In this office phase, the data were analyzed and filtered and subsequently processed. Oasis Montaj version 8.5 software from Geosof was used for this, and ArcGIS version 10.7 software from ESRI was used for charting.

For this work, the data obtained from the NOAA repository of magnetic surveys carried out in the study area between 1970 and 1971 were used as a reference. A hydrographic and bathymetric survey was performed according to technical specifications of the International Hydrographic Organization (IHO), S-44 publication for Order 2 requirements [31,32]. These regulations guarantee the quality and standardization of the results. This geophysical working method is applicable in all the deep waters of the Colombian marine territory, which are considered to be from the isobath of 100 m, to the maximum registered depth of 4600 m. This specification is also stated by Standard S44 of the IHO [31,32], which recommends that Order 2 surveys are limited to areas deeper than 100 m [33]. Nevertheless, those reference data were about 50 years old, and there were no other modern data available for the study area until the current survey. However, nautical charts 1624 and 004 edited by the CIOH in 1998 nd 2018, respectively [34,35], were taken into account to study the temporal and local variation of the geomagnetic field in the area. Thus, the declination, due to the annual variation effect, was corrected and was 4°18′ (W) in 2020. Likewise, the magnetic declination was compared with the data published in the AIP COLOMBIA Report of the Gustavo Rojas Pinilla Airport on San Andres Island, for 16 July 2020, which was 02°48′ W [36]. The difference in declinations is due to the separation between the San Andres Airport (North of the island) and the area where the magnetic declination is defined in the nautical chart 004, which is located about 90 km NE of the airport. For the analysis of magnetic declination, the procedure specified by Udias and Mezcua [23] was followed.

In this work, the Minimum Curvature Gridding or Splines method and Geographic coordinate system were used. The Gridding method refers to the process of interpolating data onto an equally spaced grid of "cells" in a specific coordinate system. This interpolation method estimates values using a mathematical function that minimizes the curvature of the surface, resulting in a smooth surface that passes exactly through the input points [37–39].

2.5. Components

Going into the field, some indispensable elements must be taken into account to carry out an optimal acquisition. For example, the magnetometry sensor and the portable winch with 300 m of telemetry cable were specifically adapted to collect the geophysical information.

The vessel ARC Providencia (Figure 5) has special adaptations, such as a winch with 2800 m of telemetry cable (Figure 6), a wet laboratory aboard the ship, and the computer center where the magnetic data are visualized and stored in real time.

Figure 5. Research vessel ARC Providencia.

Figure 6. 2800 m geophysical research winch on board the ARC Providencia.

The assembly used in the vessels is shown in Figure 7. The magnetic data are communicated from the sensor and submerged in the water, passing through the winch, the on-board cable, and finally, arriving at a "junction box" where the magnetic data are related to those of the positioning obtained by the Global Navigation Satellite System [33,40]. From there, they are transmitted to and visualized in the computer by means of MagLog software [41], as shown in Figure 8.

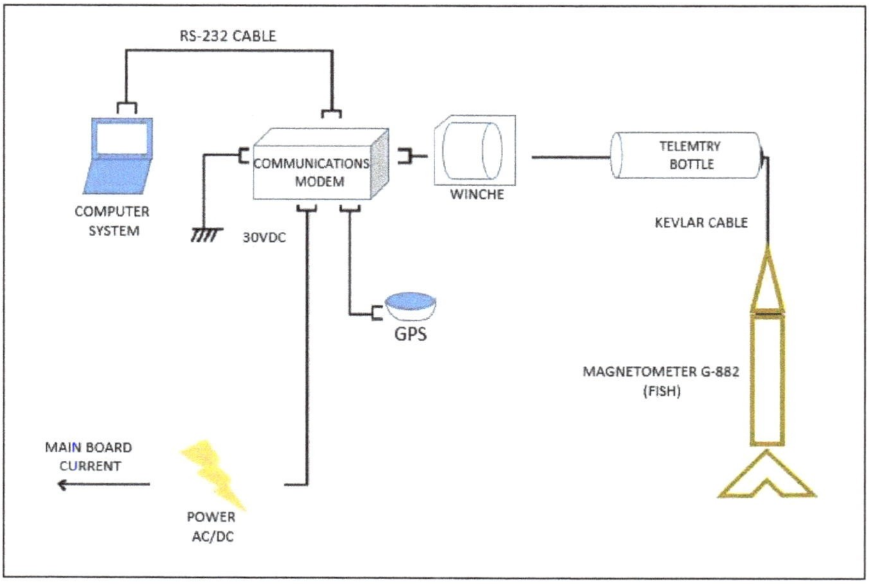

Figure 7. Assembly for the magnetic data acquisition.

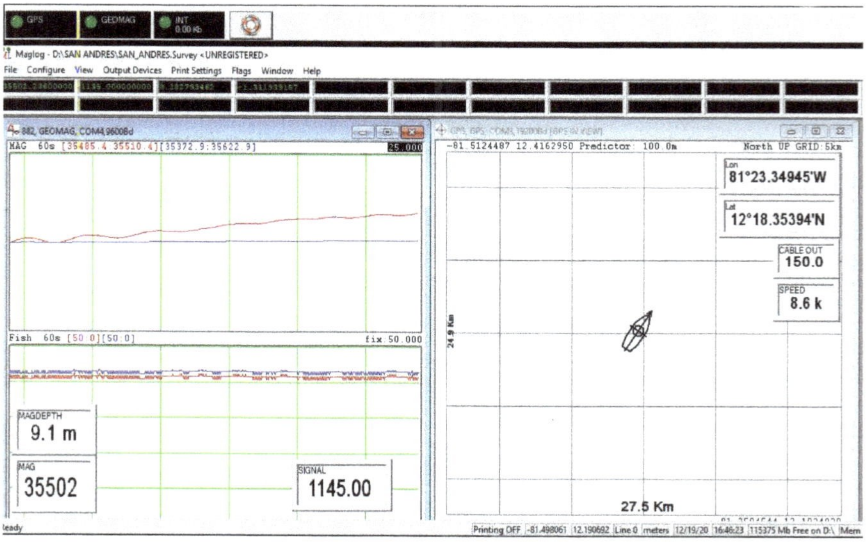

Figure 8. Visualization of geomagnetic information in MagLog software in real time.

In Figure 8, showing the visualization of data in real time, the red box to the left is the navigation window, where the position of the vessel and its course are shown. In addition, the lines that the helmsman must follow, according to the planning instructions, are indicated in this window [42]. The blue box, to the right, shows the curves of the data expected, or calculated, by the International Geomagnetic Reference Field (IGRF) model in that precise geographic position and the actual collected data. The small green indicators show the intensity of the signal, the magnetic data, and the positioning. All these indicators must be green, so that the data are correctly acquired. The red rectangle, in the extreme lower right, indicates that it is not recording, and it must be pressed to begin to record the data during the acquisition [37].

The acquisition is made with the help of the ship's personnel, taking into account certain guidelines, such as a maximum velocity of 5 knots, and a separation from the sensor of at least three times the length of the vessel, which in this case was 150 m. It is important that the "fish" is towed from the stern, as indicated in Figure 9.

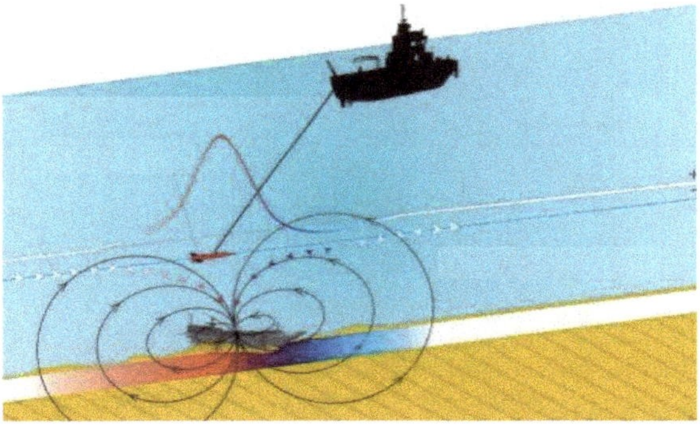

Figure 9. System of stern towing of the magnetometry equipment.

Magnetic recording is only useful in straight transects, whereas data from the turns between profile and profile are not considered, as the recorded values are affected by the magnetic field induced by the boat approaching each time a turn is made [43].

The different stages involved in the acquisition of marine geomagnetic data are subjected to a series of decisions that can radically affect the final result of the research [44]. Several usual errors exist that can be committed throughout the process, and which can be classified, according to the development stage of the study. For example, there are frequent errors related to planning that involve poor design of the lines to acquire, which could make it difficult to discern the exact form and size of the anomaly; a measurement is only of interest if the margin of error of that measurement is known. What is interpreted is a collection of data, which is why the sampling must be in accordance with the dimension of the objective to be reached. Other types of errors are associated with the measuring equipment, which can lead to mistaken readings and affect the quality of the data, operator errors, sampling errors, and errors related to environmental noise, among others. Some examples of error handling in geophysical and gravimetric data processing are shown in [45,46].

3. Results and Discussion

In the application of the method, some setbacks were presented in terms of what was planned. These were due to logistical issues with the vessel. It is also worth mentioning that the data collection was carried out on board the ship ARC Roncador, with a 300 m portable winch. The geometric arrangement had to be slightly less extensive than originally planned, as shown in Figure 10.

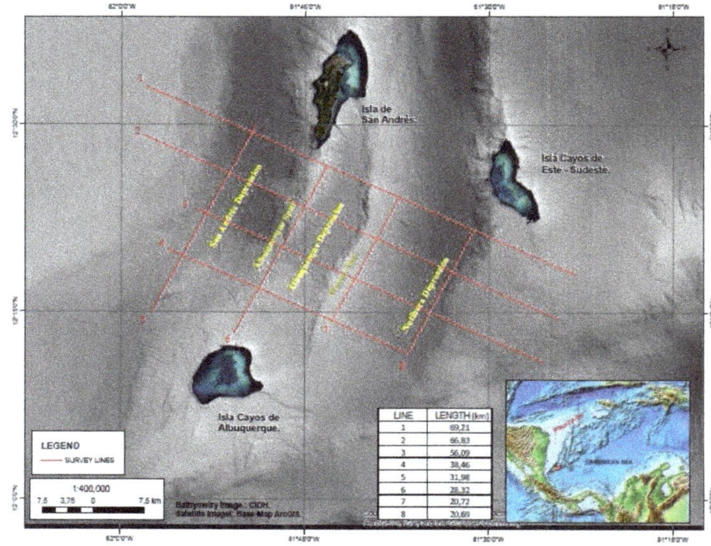

Figure 10. Geometric adjustment of the geomagnetic acquisition.

The geophysical study comprises the data collected between 20 June and 1 July 2018, in an area south of San Andrés Island, comprising four lines perpendicular to the general direction of geological structures, with a maximum length of 70.67 km, and four lines parallel to these formations, with a maximum length of 31 km. A grid-shaped geometric arrangement was preserved to provide good resolution for a regional geological study. The general direction of geological structures and geoforms has a northeast direction [47,48].

The chart of the total field of collected data appears in Figure 11, and it shows the magnetic surface of the collected data, after processing for corrections of diurnal variation, delay, direction in degrees, and of the IGRF mode [37]. A significantly positive anomaly was observed in this area, above the Nutibara Depression. The variations were in a range of −170.48 to +159.37 nT.

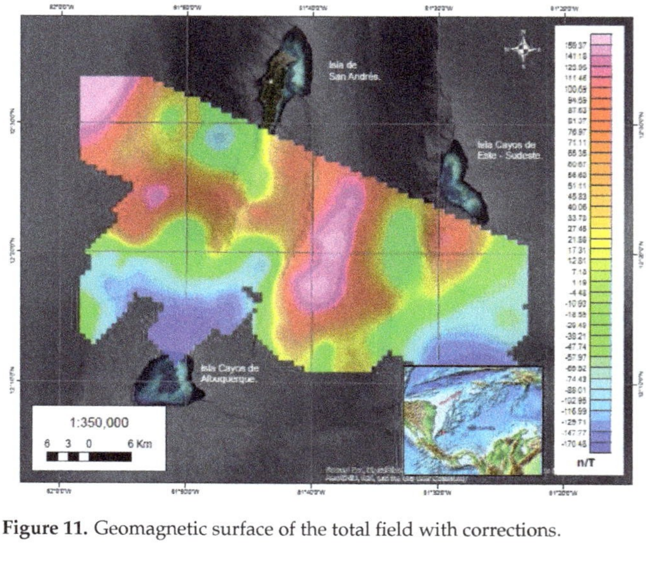

Figure 11. Geomagnetic surface of the total field with corrections.

In order to compare the obtained data with pre-existing data, a review was made of the bibliographical material and the data available from possible geomagnetic surveys carried out in the area. As a result, two research cruises were identified, giving free access to marine geomagnetic data from the National Oceanic and Atmospheric Administration [24], which are represented in two survey lines related to geophysical data that contain seismic, side-scan sonar, and magnetometry information. The first downloaded file of the zone, identified by code CH100L12, was collected by the Woods Hole Oceanographic Institution of the United States (WHOI) in 1971. The second file of the zone, identified by the code V2808, was collected by the Lamont–Doherty Earth Observatory of the United States between 1970 and 1971. The two compiled data lines appear in Figure 12.

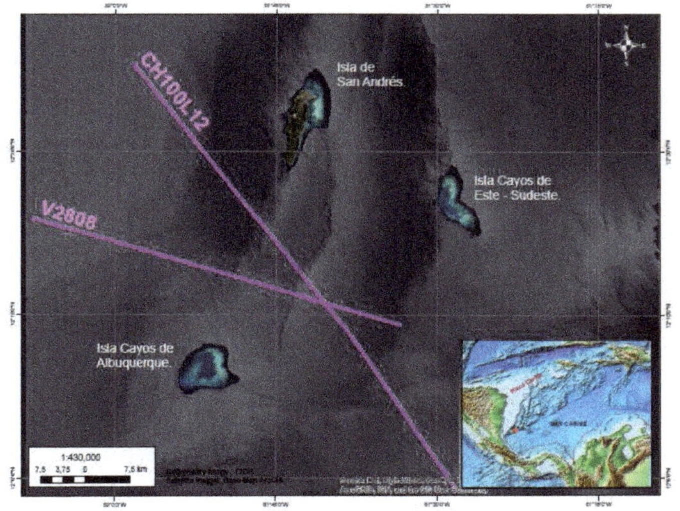

Figure 12. Tracking line chart of oceanographic cruises that acquired geomagnetic data in 1970 and 1971.

Taking into account that the magnetic field is dynamic and presents significant annual variations, an attempt was made to find geomagnetic information for the area, finding the only free access data to be those previously described, with the possible source of error of an elapsed time of around 50 years between both surveys. More current geomagnetic data on the study area have been found; however, these are global data obtained through NCEIs (National Center Environmental Information) global aeromagnetic project pertaining to the NOAA. They are aeromagnetic data obtained for the study and modeling of the Earth's magnetic field and have a much higher scale of resolution. In addition, their correction processes are different from the data obtained in situ, and specifically from those obtained in the geomagnetic surveys presented in this work. For those reasons, these aeromagnetic data were discarded for comparison, as they were incompatible in terms of resolution and processing [24]. On the other hand, the variations presented by the magnetic data during the half century between the two surveys have also been taken into account in this work. These variations have an important component of anomalies due to geological sources that have persisted in the study area during that time.

The magnetic information downloaded from NOAA [49] has an extension MGD77T and contains the positions of the tracking line and the data with corrections for diurnal variation and the IGRF. To be viewed on a common surface, the two geodatabases were joined and charted to WGS 1984 UTM Zone 17N, corresponding to the projection of the study area [39,50]. With the positional and magnetic intensity data, a magnetic surface was generated (Figure 13) showing positive and negative anomalies that vary from −328.99 to +40.48 nT.

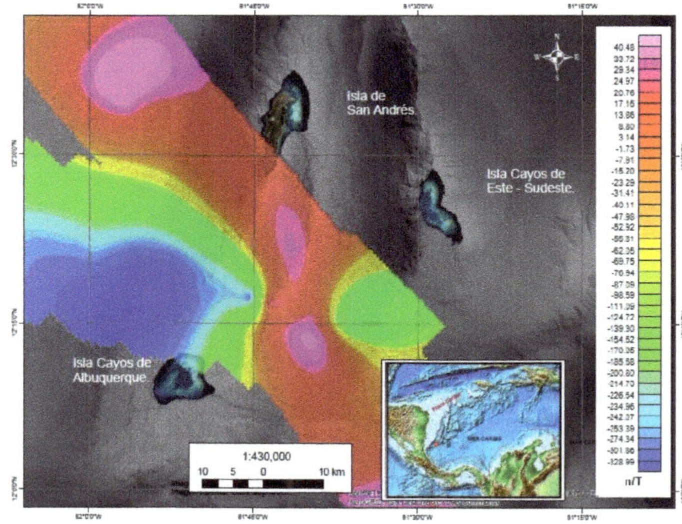

Figure 13. Geomagnetic surface of the total field, corresponding to the National Oceanic and Atmospheric Administration (NOAA) data.

In order to visualize the surface of Figure 13, a magnetic grid was generated by the Minimum Curvature Method [38], with a cell size corresponding to 1000. In this image, magenta colors are observed that are associated with positive magnetic peaks, which are located toward the northwest and over the Nutibara Depression and the Wayuu Spur; the geoforms are mentioned in Figure 10.

The layouts of the two magnetic grids are shown in Figure 14, where the positive anomalies are similarly identified on the Wayuu stimulus and the Nutibara depression, to the east and northwest, at the low Nicaraguan elevation. As for the negative anomalies, the magnetic bass located on the areas near the island of Cayos de Albuquerque stands out.

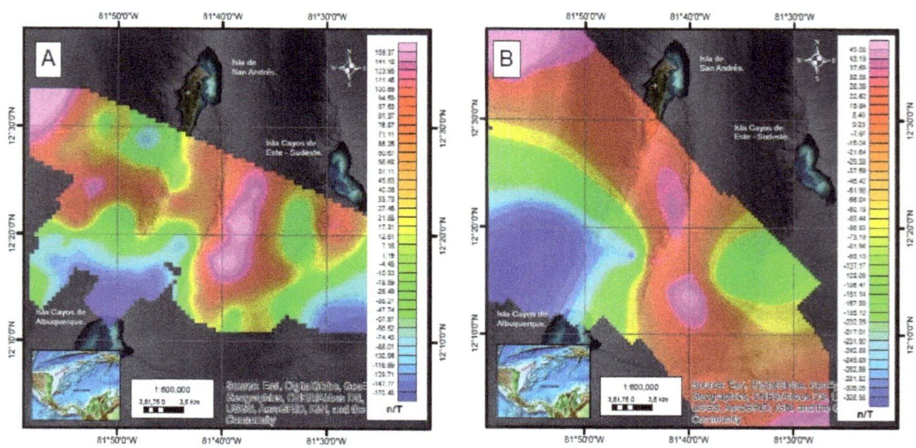

Figure 14. Geomagnetic surfaces of the total field. (**A**) Surveyed magnetic field. (**B**) Magnetic field from data downloaded from the NOAA.

It is important to mention that the color scales are not associated exactly with the same ranges on the two surfaces, but they are very close, remember that the color blue is always associated with low magnetic and pink with high magnetic [51]. These differences are due to the main factor that it is a time difference between the two surveys of around 50 years, and from which other factors that influence the acquired data can be derived, such as the accuracy of the magnetic sensor, the disposition of the field magnetic model in 1970, which presents variations with respect to that of 2018. This due to the displacement of the field and the geometric arrangement used for each case.

Comparatively speaking, in Figure 14A, the magnetic peaks are within the range of −170.48 to 159.37 nT, while in Figure 14B, the magnetic lows and highs range from −328.98 to 45.08 nT, indicating a variation of 329.85 nT in Figure 14A and 374.06 nT in Figure 14B.

Using the NOAA magnetic calculator [52] and working with the magnetic data observed for the years 1970 and 2018, small but significant differences are identified in all the measurement variables, such as the inclination with 1°46′47″, the declination with 5°2′41″, and the magnetic field represented by 4415.7 nT. These results show a variation of the magnetic field across the timeline. It is important to highlight that these anomalies identified in the study area are geological in nature, which infers that they will be present as sources of magnetic anomalies for a very long period of time.

The significant magnetic anomalies, which can be seen in Figure 11, marked in fuchsia/magenta, correspond to a magnetic high. In Figure 10, it can be seen that this anomaly corresponds to the geomorphology of the Nutibara depression, and that it could be generated by some type of mineral deposition that can be found, associated with ferrous materials, or also with volcanic material with a high iron content compared to its geological environment. On the other hand, magnetic lows are observed north of the Cayos de Alburquerque Island, which would seem to be a contradiction, since its morphology is typical of a seamount, and, therefore, its magnetic response should be high [48,53,54].

The results of this work are very important for the scientific community, as this is an area where there is an immense lack of data. The data obtained in the survey, carried out in 2017 by the CIOH, 50 years later, are very important data taken in situ, with a high resolution that make them very reliable and precise, and they are unique in the area of the Archipelago of San Andres, Providencia and Santa Catalina.

We have carried out an exhaustive search for data and geomagnetic surveys in the study area and, although no new works have been found, similar or related works have been carried out in the environment of the study area: magnetic mapping of the northern Caribbean region using marine magnetic data from GEODAS [55], works about gravity and magnetic field referring to hydrocarbon prospects at the Tobago Basin [56], geological description and interpretation in Providencia and Santa Catalina Islands [57], and many others related to tectonics and volcanism [58,59].

4. Conclusions

Surveying for the acquisition of geomagnetic data in marine environments is a method that is gaining ground worldwide, and it offers great opportunities for development and advances in new lines of research and scientific knowledge. In addition, it is a technology that supports different research and engineering projects, such as the detection of the location of pipelines and covers, buried ordnance, shipwrecks, identification of sites of archaeological interest, and the characterization of geological structures, among other applications, and also in projects related to national sovereignty and the study of a country's natural resources.

The methodology for marine geomagnetic acquisition has become the prime standard for marine geophysical research for the study of national resources and Colombian sovereignty.

Although geophysical exploration is dominated by established foreign companies, DIMAR now has the capacity to offer geophysical magnetometry services to different countries within its sphere, with excellent technical and human resources, and research equipment and vessels.

After much effort and field tests in the Colombian Caribbean, the geomagnetic acquisition procedure has been standardized as a methodology that can obtain high-quality marine geomagnetic information.

The acquisition of the G-882 marine magnetometer, the application of this methodology to a survey in the Colombian Caribbean, and the development of the magnetometry method have responded to the need of the DIMAR to recover the capacity for scientific research at national level, and scientific leadership in the region, by having an efficient tool in geological and archaeological prospecting supported by two modern, well-equipped scientific research platforms, namely, the ARC Malpelo and ARC Providencia research vessels.

Small differences have been identified between the magnetic data obtained for the years 1970 and 2018, being negligible in the variables measured, such as the inclination, declination, and total magnetic field. These results show a variation of the magnetic field across the very long timeline, so it can be inferred that these anomalies in the study area have an important geological component and will be present for a long time. These differences may also be attributable to the acquisition and processing methods used in the 1970s.

The results of this work are very important for the scientific community, because this is an area where there is a great lack of magnetic data. The data from the survey carried out in 2017 by the CIOH are very important due to the survey resolution reached, having achieved 91,285 data taken in the field at a rate of 20 data/second with an average ship speed of 7 knots, which managed to obtain a datum every 2 cm (0.02 m/datum).

Author Contributions: Conceptualisation, B.J.A.; J.J.M.-P. and K.O.P.; investigation and writing—original draft preparation, K.O.P., B.J.A., J.J.M.-P., J.R.C., N.O.M., F.C.-d.-V.; review and editing B.J.A., J.J.M.-P., F.C.-d.-V. All authors have read and agreed to the published version of the manuscript.

Funding: The research for the acquisition and capture of geomagnetic data was supported by the Caribbean Oceanographic and Hydrographic Research Center (CIOH), attached to the General Maritime Directorate within the framework of the "Marine Geomagnetism Project". The APC was funded by CIOH and RNM912 Coastal Engineering Research Group of the University of Cadiz.

Data Availability Statement: The data presented in this study are available on request from the Centro de Investigaciones Oceanograficas e Hidrograficas de Colombia. The data are not publicly available due to military restrictions.

Acknowledgments: This work was possible thanks to the support of the "Marine Geomagnetism" project, financed by the General Maritime Directorate. The authors thank the crew of the ARC Roncador Oceanographic Research Vessel and the staff of the survey area of the Colombian Hydrographic and Oceanographic Research Center (CIOH) for their collaboration during the survey campaigns. We also thank the Captain of the Navy, Germán Augusto Escobar Olaya, Director General of CIOH for his support in the fieldwork and authorisation for the use of CIOH data for the preparation of this document. The authors thank Javier Idárraga García, the editors and the two anonymous reviewers for their comments and suggestions which greatly improved the manuscript.

Conflicts of Interest: The authors declare no conflict of interest.

References

1. Allan, T. A review of marine geomagnetism. *Earth-Sci. Rev.* **2003**, *5*, 217–254. [CrossRef]
2. Ewing, M.; Heezen, B.C.; Ericson, D.B.; Northrop, J.; Dorman, J. Exploration of the Northwest Atlantic mid-ocean canyon. *GSA Bull.* **1953**, *64*, 865–868. [CrossRef]
3. Campbell, W.H. *Introduction to Geomagnetic Fields*, 2nd ed.; Cambridge University Press: Cambridge, UK, 2003; p. 337. ISBN 9781139165136.
4. Roy, K.K. *Potential Theory in Applied Geophysics*; Springer: Berlin/Heidelberg, Germany, 2008; p. 651. ISBN 13:9783540720898.
5. MnDOT (Minnesota Department of Transportation). Report Minnesota Deep Test Protocol Project, Chapter 11: COMPARISON OF OUTCOMES AND COSTS OF METHODS: FOUNDATIONS FOR A DEEP TEST PROTOCOL, Mn/DOT/WR-0200. Available online: https://www.dot.state.mn.us/culturalresources/docs/deepsite/11CostBenefitScreen.pdf (accessed on 1 July 2020).
6. Gallego, J.M.; Jaramillo, H.; Patiño, A. Servicios Intensivos en Conocimiento en la Industria del Petróleo en Colombia. Banco Interamericano de Desarrollo (BID), Documento para Discusión N° IDB-DP-417. 2015. Available online: https://pdfs.semanticscholar.org/b227/56f3b46037312b51e7dc93899ecbb1e0bb29.pdf (accessed on 10 October 2020).

7. Jigena, B.; Berrocoso, M.; Torrecillas, C.; Vidal, J.; Barbero, I.; Fernandez-Ros, A. Determination of an experimental geoid at Deception Island, South Shetland Islands, Antarctica. *Antarct. Sci.* **2016**, *28*, 277–292. [CrossRef]
8. Muñoz-Pérez, J.J.; Gutiérrez, J.M.; Naranjo, J.M.; Torres, E.; Fages, L. Position and monitoring of anti-trawling reefs in the Cape of Trafalgar (Gulf of Cadiz, SW Spain). *Bull. Mar. Sci.* **2000**, *67*, 761–772.
9. DIMAR (Dirección General Marítima de Colombia). *Manual de Adquisición y Procesamiento de Información Geomagnética*; DIMAR: Cartagena de Indias, Colombia, 2015; p. 180.
10. Milsom, J.; Eriksen, A. Field geophysics, fourth edition. *Environ. Eng. Geosci.* **2013**, *19*, 205–206. [CrossRef]
11. Schmidt, A. Archeology, magnetic methods. In *Encyclopedia of Geomagnetism and Paleomagnetism*; Gubbins, D., Herrero-Bervera, E., Eds.; Encyclopedia of Earth Sciences Series; Springer: Berlin/Heidelberg, Germany; New York, NY, USA, 2007; pp. 23–31. ISBN 978-1-4020-3992-8. [CrossRef]
12. Rodríguez Álvarez, R. Geofísica General. 2010. Available online: https://www.medellin.unal.edu.co/~{}rrodriguez/geologia/geofisica.htm (accessed on 30 June 2020).
13. Deiana, R.; Leucci, G.; Martorana, R. New perspectives on geophysics for archaeology: A special issue. *Surv. Geophys.* **2018**, *39*, 1035–1038. [CrossRef]
14. Pérez-Orrego, C.E. Modelamiento e Interpretación de Datos Magnéticos y Gravimétricos en la Caracterización de un Depósito Tipo IOGC. Master's Thesis, Facultad de Ciencias Físicas y Matemáticas, Departamento de Geofísica, Universidad de Chile, Santiago, Región Metropolitana, Chile, 2017; p. 127. Available online: http://repositorio.uchile.cl/bitstream/handle/2250/149082/Modelamiento-e-interpretacion-de-datos-magneticos-y-gravimetricos-en-la-caracterizacion-de-un-deposito.pdf?sequence=1&isAllowed=y (accessed on 8 July 2020).
15. Tchernychev, M.; Johnson, R.; Johnson, J.; Geometrics Inc. Multiple Magnometer Sensor Array and Their Application in Marine Environmental Surveys. Ocean News & Technology. May 2014. Available online: https://www.oceannews.com/featured-stories/may-multiple-magnetometer-sensor-arrays-and-their-applications-in-marine-environmental-surveys (accessed on 17 July 2020).
16. Telford, W.M.; Gerald, L.; Sheriff, R. *Applied Geophysics*, 2nd ed.; Cambridge University Press: Cambridge, UK, 1990; p. 760.
17. Ariffin, K.S. Geophysical Surveying Using Magnetics Methods. EBS 309: Geofiz Carigali, Magnetik. Adapted from ES304—Geophysical Prospecting, Earth Science Department University of Melbourne, Australia, n.d. Available online: http://www.ukm.my/rahim/G-Magnetics%20Methods.pdf (accessed on 20 August 2020).
18. Goguitchaichvilia, A.; Hernández-Quintero, E.; García, R.; Cejudo, R.; Cifuentes, G.; Cervantes, M. Fluctuation of the Earth's magnetic field elements in Mexico revealed by archive documents since 1587. *Phys. Earth Planet. Inter.* **2020**, *300*, 106433. [CrossRef]
19. Ramirez, E.; Sánchez-Rojas, J.; Orihuela, N.; Schmitz, M. Adquisición y procesamiento de datos gravimétricos y magnéticos en el Perfil Calabozo. In Proceedings of the Jornadas de Investigación Encuentro Académico Industrial, Universidad Central de Venezuela, Dabajuro, Caracas, Venezuela, 25–28 November 2014. Available online: https://www.researchgate.net/publication/283488360_ADQUISICION_Y_PROCESAMIENTO_DE_DATOS_GRAVIMETRICOS_Y_MAGNETICOS_EN_EL_PERFIL_CALABOZO_-_DABAJURO (accessed on 16 November 2020).
20. Hatakeyama, T.; Kitahara, Y.; Yokoyama, S.; Kameda, S.; Shiraishi, J.; Tokusawa, K.; Mochizuki, N. Magnetic survey of archaeological kiln sites with Overhauser magnetometer: A case study of buried Sue ware kilns in Japan. *J. Archaeol. Sci. Rep.* **2018**, *18*, 568–576. [CrossRef]
21. Hinze, W.J.; Von Frese, R.R.B.; Saad, A.H. *Gravity and magnetic exploration: Principles, Practices, and Applications*; Cambridge University Press: Cambridge, UK, 2017; p. 502. ISBN 978-0-521-87101-3. Available online: https://assets.cambridge.org/97805218/71013/frontmatter/9780521871013_frontmatter.pdf (accessed on 8 July 2020).
22. Reeves, C. *Aeromagnetic Surveys: Principles, Practice and Interpretation*; Geosoft: Delft, The Netherlands, 2005; Volume 35. Available online: https://files.seequent.com/PDFs/Aeromagnetic_Survey_Reeves-october-2005.pdf (accessed on 29 July 2020).
23. Udías Vallina, A.; Mezcua Rodriguez, J. *Fundamentos de Geofísica*; Alianza Editorial: Madrid, Spain, 1997; p. 480. ISBN 978-84-206-8167-2.
24. NOAA-NCEI (NOAA National Centers for Environmental Information, NCEI). Marine Geology and Geophysics > Geophysics. Available online: https://www.ngdc.noaa.gov/geomag/geomag.shtml (accessed on 30 June 2020).
25. Fernández-Montblanc, T.; Quinn, R.; Izquierdo, A.; Bethencourt, M. Evolution of a Shallow Water Wave-Dominated Shipwreck Site: Fougueux (1805), Gulf of Cadiz. *Geoarchaeology* **2016**, *31*, 487–505. Available online: https://onlinelibrary.wiley.com/doi/pdf/10.1002/gea.21565 (accessed on 17 May 2020). [CrossRef]
26. Fernández-Montblanc, T.; Del Río, L.; Izquierdo, A.; Gracia, F.; Bethencourt, M.; Benavente, J. Shipwrecks and man-made coastal structures as indicators of historical shoreline position. An interdisciplinary study in the Sancti Petri sand spit (Bay of Cádiz, SW Spain). *Mar. Geol.* **2018**, *395*, 152–167. [CrossRef]
27. Introcaso, A.; Ghidella, M.E.; Ruiz, F.; Crovetto, C.B.; Introcaso, B.; Paterlini, C.M. Métodos gravi-magnetométricos modernos para analizar las características estructurales de la plataforma continental argentina. *Geoacta* **2008**, *33*, 1–20.
28. Plets, R.; Dix, J.; Bates, R. *Marine Geophysics Data Acquisition, Processing and Interpretation. Guidance Notes*; Jones, D.M., Ed.; English Heritage: Swindon, UK, May 2013; 48p, Available online: https://historicengland.org.uk/images-books/publications/marine-geophysics-data-acquisition-processing-interpretation/mgdapai-guidance-notes/ (accessed on 20 November 2020).

29. Barrows, L.; Rocchio, J.E. Magnetic surveying for buried metallic objects. *Ground Water Monit. Remediat.* **1990**, *10*, 204–211. [CrossRef]
30. Geometrics. G−882 CESIUM MARINE MAGNETOMETER 25919-OM REV. FI. Operation Manual, Copyright. 2005. Available online: https://geometrics.com/wp-content/uploads/2018/10/G-882Man_RevF.pdf (accessed on 20 June 2020).
31. IHO (International Hydrographic Organization). *IHO Standards for Hydrographic Surveys*, 6th ed.; IHO: Monte Carlo, 2020. Available online: https://iho.int/uploads/user/pubs/Drafts/S-44_Edition_6.0.0-Final.pdf (accessed on 15 November 2020).
32. Carrillo de Albornoz, P.; Francisco, J. El Instituto Hidrográfico de la Marina. Revista Arbor, Consejo Superior de Investigaciones Científicas CSIC, Arbor CLXXIII, 682. Octobre 2002, pp. 365–383. Available online: https://core.ac.uk/download/pdf/207331548.pdf (accessed on 15 October 2020).
33. Jigena, B.; de Gil, A.; Walliser, J.; Vidal, J.; Muñoz-Perez, J.J.; Pozo, L.; Lebrato, J. Improving the learning process in the subject of basic maritime training using GPS and Google Earth as useful tools. In Proceedings of the INTED2016 Conference, Valencia, Spain, 7–9 March 2016; pp. 6161–6171. ISBN 978-84-608-5617-7.
34. CIOH (Centro de Investigaciones Oceanograficas e Hidrograficas). Carta Número 1624, Islas de San Andrés y Providencia, 1ra. Ed. Enero 1998, Escala 1:100.000. 1998. Available online: https://www.cioh.org.co/derrotero/wrapperDerrotero/cartas/carta_papel.php?codigocarta=1624 (accessed on 16 October 2020).
35. CIOH (Centro de Investigaciones Oceanograficas e Hidrogrtaficas). Carta Número 004, Archipielago de San Andres y Providencia, 3a Ed. 2018, Escala 1: 600000. 2018. Available online: https://www.cioh.org.co/derrotero/wrapperDerrotero/cartas/carta_papel.php?codigocarta=004 (accessed on 16 October 2020).
36. AIP COLOMBIA (Servicios de Información Aeronáutica de la Dirección de los Servicios a la Navegación Aérea). *Informe AIP AD 2—SKSP 1 del 16 de Julio de 2020*; Aeródromo SKSP: San Andreas, Colombia, 2020.
37. Tchernychev, M. MAGPICK-Magnetic Map&Profile Processing User Guide. Beta Version of the Manual—Last Revised 12/12/2013, Copyright (C) 1997 Mikhail Tchernychev. Available online: https://geometrics.com/wp-content/uploads/2018/10/MagPick.pdf (accessed on 15 July 2020).
38. Geosoft. Oasis Montaj How-to Guide. Oasis Montaj Gridding. Available online: http://updates.geosoft.com/downloads/files/how-to-guides/Oasis_montaj_Gridding.pdf (accessed on 15 July 2020).
39. Fernández-Coppel, I.A. Las Coordenadas Geográficas y la Proyección UTM. Universidad de Valladolid, Escuela Técnica Superior de Ingenierías Agrarias, Palencia. 2001. Available online: https://www.cartesia.org/data/apuntes/cartografia/cartografia-geograficas-utm-datum.pdf (accessed on 20 August 2020).
40. Geometrics. G-856 Memory-MagTM Proton Precession Magnetometer OPERATOR'S MANUAL. P/N 18101-02. Available online: http://userpage.fu-berlin.de/~{}geodyn/instruments/Manual_Geometrics_G856.pdf (accessed on 16 November 2020).
41. Geometrics. MagLogPro™ MagLogLite™ Data Acquisition Software 25479-01 Rev. R. Available online: https://geometrics.com/wp-content/uploads/2018/10/MagLogManual.pdf (accessed on 20 June 2020).
42. Schwartz, A.B. Magnetometer system for detecting OE in very shallow marine environments. In *Symposium on the Application of Geophysics to Engineering and Environmental Problems*; 2004; pp. 1722–1733. ISSN 1554-8015. Available online: https://doi.org/10.4133/1.2923318 (accessed on 20 November 2020).
43. Calcina, M. UN MODELO DINÁMICO PARA EL CAMPO GEOMAGNÉTICO. Revista Boliviana de Física. Cited on 20 June 2020. Volume 15, pp. 44–62. Available online: http://www.scielo.org.bo/scielo.php?script=sci_arttext&pid=S1562-38232009000100008&lng=es&tlng=es (accessed on 20 November 2020).
44. Geotem Ingeniería, S.A. Magnometría Marina. Available online: http://www.geotem.com.mx/marinos4.php (accessed on 9 June 2020).
45. Berrocoso, M.; Prates, G.; Fernández-Ros, A.; Peci, L.M.; De Gil, A.; Rosado, B.; Páez, R.; Jigena, B. Caldera unrest detected with seawater temperature anomalies at Deception Island, Antarctic Peninsula. *Bull. Volcanol.* **2018**, *80*, 41. [CrossRef]
46. Berrocoso, M.; Gárate, J.; Martín, J.; Fernández, A.; Moreu, G.; Jigena, B. Improving the local geoid with GPS. In *Reports of the Finnish Geodetic Institute 96.2*; Finnish Geodetic Institute: Helsinki, Finland, 1996; pp. 96–102.
47. Javier Idárraga-García, J.; Masson D., G.; García, J.; León, H.; Vargas, C.A. Architecture and development of the Magdalena Submarine Fan (southwestern Caribbean). *Mar. Geol.* **2019**, *414*, 18–33. [CrossRef]
48. Idárraga-García, J.; León, H. Unraveling the underwater morphological features of roncador bank, archipelago of San Andres, Providencia and Santa Catalina (Colombian Caribbean). *Front. Mar. Sci.* **2019**, *6*. [CrossRef]
49. NOAA-NCEI (NOAA National Centers for Environmental Information, NCEI). Bathymetric Data Viewer. Available online: https://maps.ngdc.noaa.gov/viewers/bathymetry/ (accessed on 30 June 2020).
50. Berrocoso, M.; Paez, R.; Jigena, B.; Caturla, C. The RAP net: A geodetic positioning network for Andalusia (South Spain). In Proceedings of the EUREF Mitteilungen des Bundesamtes für Kartographie und Geodäsie, Riga, Latvia, 14–17 June 2006; pp. 364–368.
51. Narváez Medina, L. Modelo de Fuentes de Anomalías Geomagnéticas de Campo Total Asociadas al Estado de la Cámara Magmática del Volcán Galeras. Universidad Nacional de Colombia, 2012. Available online: https://repositorio.unal.edu.co/bitstream/handle/unal/75128/194379.2013.pdf?sequence=1&isAllowed=y (accessed on 15 December 2020).
52. NOAA-NCEI (NOAA National Centers for Environmental Information, NCEI). Magnetic Field Calculators. Available online: https://www.ngdc.noaa.gov/geomag/calculators/magcalc.shtml#igrfwmm (accessed on 30 June 2020).
53. Milliman, J.D.; Supko, P.R. On the geology of San Andres Island, western Caribbean. *Geol. Mijn.* **1968**, *47*, 102–105.

54. Geister, J.; Díaz, J.M. Reef Environments and Geology of an Oceanic Archipelago: San Andres, Old Providence and Sta. Catalina (Caribbean Sea, Colombia). Colombia: Boletín Geológico Instituto Nacional de Investigaciones Geológico Mineras, 142. 2007. Available online: https://www2.sgc.gov.co/Publicaciones/Cientificas/NoSeriadas/Documents/AmbGeolArch.SAnd.Prov.SCat.en.PDF (accessed on 20 November 2020).
55. Catalán, M.; Martín-Dávila, J.J. Lithospheric magnetic mapping of the northern Caribbean region. *Geol. Acta* **2013**, *1*, 311–320.
56. Yuan, B.; Song, L.; Han, L.; An, S.; Zhang, C.; Bingqiang, Y.; Li, H.; Shaole, A.; Chunguan, Z. Gravity and magnetic field characteristics and hydrocarbon prospects of the Tobago Basin. *Geophys. Prospect.* **2018**, *66*, 1586–1601. [CrossRef]
57. Alvarez-Gutiérrez, Y.; Amaya-López, C.; Barbosa-Mejía, L.N.; Builes-Carvajal, J.S.; Henao-Casas, J.D.; Montoya-Cañola, S.M.; Pacheco-Sintura, P.A.; Ramírez-Hoyos, L.F.; Urrego-Osorio, S.; Zapata-Montoya, A.M.; et al. Geological description and interpretation of Providencia and Santa Catalina Islands. *Boletín Cienc. Tierra* **2014**, *35*, 67–81.
58. Xu, X.; Keller, G.R.; Guo, X. Dip variations of the North American and North Caribbean Plates dominate the tectonic activity of Puerto Rico-Virgin Islands and adjacent areas. *Geol. J.* **2015**, *51*, 901–914. [CrossRef]
59. Geldmacher, J.; Hoernle, K.; Bogaard, P.V.D.; Hauff, F.; Klügel, A. Age and Geochemistry of the Central American Forearc Basement (DSDP Leg 67 and 84): Insights into Mesozoic Arc Volcanism and Seamount Accretion on the Fringe of the Caribbean LIP. *J. Pet.* **2008**, *49*, 1781–1815. [CrossRef]

Journal of
Marine Science and Engineering

MDPI

Article

Multidata Study to Evaluate the Impact of Submarine Outfall in a Beach Sedimentary Dynamic: The Case of Samil Beach (Galicia, Spain)

Aimar Lersundi-Kanpistegi [1], Ana M. Bernabeu [1,*], Daniel Rey [1] and Rafael Díaz [2]

[1] Centro de Investigación Mariña, Universidade de Vigo, GEOMA, Edificio Olimpia Valencia, Campus de Vigo, As Lagoas, Marcosende, 36310 Vigo, Spain; adelvalle@uvigo.es (A.L.-K.); danirey@uvigo.es (D.R.)

[2] Acuaes, delegación Vigo, 36201 Vigo, Spain; rafael.diaz@acuaes.com

* Correspondence: bernabeu@uvigo.es

Received: 8 June 2020; Accepted: 18 June 2020; Published: 23 June 2020

check for updates

Abstract: The Ria de Vigo (NW Iberian Peninsula) is one of the most impacted coastal areas of Galicia, due to demographic and industrial pressure. One of the main consequences of this pressure is the need to extend the current wastewater treatment plant of the city of Vigo (295,000 inhabitants). This extension includes a new submerged pipeline construction to discharge the treated water in the central channel of the Ria. The new planned pipeline must cross Samil Beach, the most important urban beach of the city. Based on a multitool strategy, this work characterizes the interactions between the new pipeline route alternatives and the sediment dynamics of Samil Beach. This approximation improves the reliability of the results in the subtidal area of the beach, where studies are scarce due to the complexity of the data acquisition. The present study is based on high resolution bathymetry data, seabed physical characterization, a granulometric study of the superficial sediment, and a numerical simulation of the tide, wave climate, and sediment transport in low and high energy conditions using open source Delft3D software. The results showed that the area of interest is a low energy area, which is significantly shielded from wave attack, where fine sand predominates. However, the field data indicated an interaction (accretion-erosion) in the submerged obstacles between 0 and 12 m deep. The model revealed that there is significant sediment movement above a 7.4 m isobath, and that the pipeline would not alter the general transport dynamics of the beach, but would interact in the shallowest section. The main conclusion of this work states that the future structure would not alter the global sediment dynamics of the beach. In addition, in order to guarantee the safety of the new pipeline, it should emerge above an 8 m isobath. The multiapproach methodology presented can be applied to other studies of the interaction between coastal structures and the environment.

Keywords: beach evaluation; multidata approach; sedimentary dynamics; outfall

1. Introduction

At present, shores worldwide undergo very high pressure due to their inherent touristic interest and industrial development favored by access to the sea, and the effect of climate change [1–4]. A deficient management of coastal activities has led to many environmental problems, such as serious morphologic coastline alterations, pollution, and the loss of important habitats for coastal ecosystems and the local economy. Therefore, it is essential to have a good knowledge of the coastal environment in order to guarantee an appropriate management to preserve both environment and economic activities [5–7].

In this context, the appropriate design of coastal structures is central, since they are indispensable for socioeconomic development, but may cause serious alterations in the coastal environment. According to the Recommendations for Maritime Works collection (RMW), published by Puertos del Estado (http://www.puertos.es/es-es/ROM; Spanish Ministry of Public Works), these structures must simultaneously satisfy three requirements; security, service and exploitation. The security requirement states that the design of the construction must guarantee the physical integrity of the structure itself. The service requirement states that the design must guarantee the service for which the structure is built, and the exploitation requirement states that it must be exploited without serious repercussions in the socioeconomic framework and environment.

From the geotechnical and environmental point of view, one of the requirements in coastal projects is the study of the interaction produced between the structure and the physical environment. These interferences are produced by physical agents (wind, waves, and currents) due to the presence of the structure, and must be predicted in the design; otherwise, it may lead to serious structural problems, and failings in one or more RMW requirements. Accordingly, it is very difficult to predict the interaction between the structure and the local sedimentary dynamics, a process that can have serious consequences both for the environment and the structure itself [8]. From the environmental point of view, the lack of forecast can cause habitat and water quality losses, and permanent morphological alterations [3]. Economically, this lack can also lead to marine (i.e., fisheries) and touristic resource loss, which has a negative social impact. Finally, it can also produce the total failure of the structure functionality. Therefore, it is of great importance to diminish to minimum levels the uncertainties in the forecast of the interaction between the structure and the environment during the design process.

Several methodological approaches have been proposed for the study of the structure—environment interaction [9,10]. Some studies have used photographic interpretation and Geographic Information System (GIS) methods to evaluate the coastal evolution [11]. Other authors used beach and oceanographic parameters monitoring [12,13]. Some studies are based on numerical modeling to study the morphological and sediment transport impact of the coastal structures on beaches [14–17].

The water-treatment plant E.D.A.R.-Lagares treats the sewage of the Vigo (295,000 inhabitants) coastal city, and empties into the central channel of the Ria de Vigo at a 35 m depth. This plant was built in 1997, with a treatment capacity of 400,000 inhabitants equivalent (IE). Nevertheless, the increase of 14,000 inhabitants in 11 years (1998–2009), and the entry into the force of the E.U. Water Framework Directive necessitated the extension and modernization of the plant. The new treatment station has an 800,000 IE capacity, and treats the storm runoff waters without directly emptying into the Ria during large rain events. This extension implied the construction of a new pipeline larger than the two current ones, which must cross the Samil Beach until the central channel bottom of the Ria is reached, at a 35 m depth.

The main purpose of this work was to determine and characterize the interactions that can be produced between the new pipeline route alternatives and the sediment dynamics of Samil Beach, providing greater reliability to the project. This research was performed using a multitool strategy in order to improve the results, given the complex nature of the environment, and to support the decision-making in the first steps of the construction project, allowing for the mitigation of beach erosion and the protection of valuable infrastructure.

2. Study Area

The Ria de Vigo is the southernmost coastal inlet of the Rias Baixas. It presents a funnel shape in plan view, and its axis is oriented NE–SW (head-mouth). The Ria is 30 km long and 0.6 km wide in its inner part, and 15 km wide at the outer part. This inlet presents rocky islands at the entrance, Cíes Islands, and a central deep channel that reaches a 50 m depth in the southern mouth.

Samil Beach is located in the southern margin of the outer sector of the Ria de Vigo (Figure 1), in a small bay oriented to the NW. The bay is bounded in the SW by the Toralla Island and the large rocky outcrop of Estai Cape-Canido, and in the east by the Mar Cape. Samil Beach is 1.9 km long, occupying half of the bay coast. This beach presents reflective morphodynamic elements and fine sand (209 μm), typical of the middle Ria sector, based on exposure to the open sea waves and sediment provenance [18].

The seabed in the southern margin of the Ria is dominated by sands, with the presence of gravels [19]. Samil Beach is separated by a rocky outcrop from the Vao Beach, and in their submerged sector there are also outcrops presenting magnitudes in the order of tens of meters.

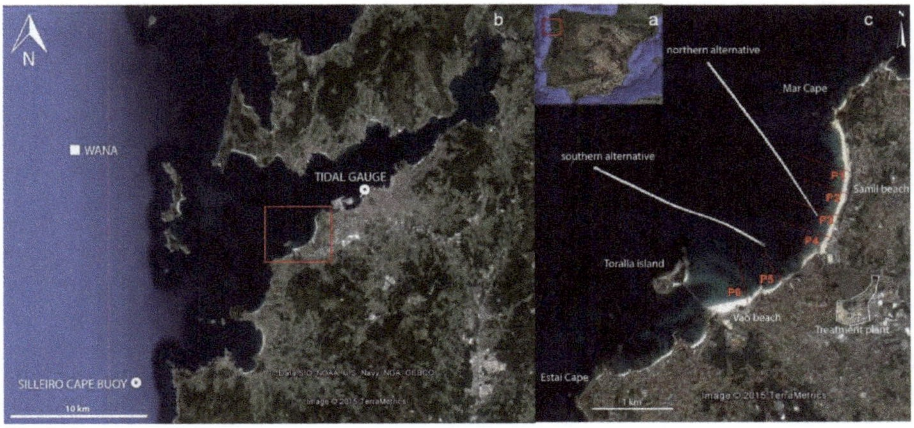

Figure 1. Location of the study area: (**a**) location of Rias Baixas in Spain; (**b**) the Ria de Vigo showing the Silleiro Cape buoy, WANA model node, and the tidal gauge of the Vigo harbour; (**c**) satellite photography (Google Earth©) of the area of interest, showing the two alternative routes designed for the submarine pipeline of the wastewater treatment plant (in white) and beach profiles (in red) considered in the seasonal dynamics analysis.

The currents in the Rias Baixas are controlled by the tide, the fluvial discharge and the shelf wind pattern. The tide presents a semidiurnal and mesotidal regime, with a mean range of 2.2–2.4 m [20,21], a mean velocity of 5–10 cm/s and a maximum velocity of up to 30 cm/s in the shallowest and narrowest sections [22]. The fluvial discharges in the Rias Baixas are seasonal; they are at their maximum in winter and minimum in summer [23]. These discharges are in general of minor importance, given the size of their drainage basins (620–3600 km^2) compared to the area of the Rias (106–251 km^2). The shelf wind action produces the deep-water upwelling events in the coast under the north winds, prevailing between March and October, and it causes a downwelling of superficial waters under south winds, which prevail for the rest of the year. Such currents are subtidal and show a smaller magnitude than the tidal ones (<6 cm/s).

The wave climate in this region also presents strong seasonal behaviour. In summer, the significant wave height (Hs) is 1–1.5 m, with a peak period (Tp) of 8–10 s, whilst the winter shows higher energy conditions, with an Hs of 1.5–3 m and a Tp of 12–14 s [24]. In extreme conditions, the Hs can reach values of 8 m [25,26]. The predominant incoming direction is WNW–NNW, presenting a 76.3% annual frequency. The SW waves are not very frequent in the region, at 8.0%, but they are usually linked to high energy conditions [27]. Inside the Rias, the wave conditions are modified significantly depending on the incoming wave direction due to the presence of the Cíes Islands and the Ria axis orientation. Therefore, the offshore waves undergo lower changes and easily reach the inner zones of the Ria when their direction matches the orientations of the mouths, i.e., the NW and SW waves. In addition, SW wave events coincide with the natural orientation of the Rias reaching the innermost areas. In addition, such wave events are generally related to winter storms, and their effects in the bottom may be important.

Characteristics and Alternative Routes of the Submarine Pipeline

The new pipeline consists of a high-density plastic pipe with an inner diameter of 1.8 m. In the first section of the route, the pipeline is underground, and it emerges gradually on the seabed from a 3.3 m isobath. From this point, the pipeline is protected by a concrete structure with a trapezoidal section, which increases its size gradually with the depth until it reaches a 20 m base width, a 4 m crest width and a height of 5 m from the bottom.

Two possible routes were designed according to the bay morphology (Figure 1). In the southern alternative, the pipeline departed from the southern sector of the bay, specifically from the rocky outcrop located between the Samil and Vao Beaches, with a direction of 300° (with respect to the north clockwise). In the northern alternative, the pipeline departed from the central sector of the bay, close to the current submerged pipelines, with a direction of 340° (with respect to the north clockwise).

3. Methodology

3.1. Field Work

The bathymetry of the subtidal area of Samil Beach was acquired in July 2011 up to a 38 m depth with a Reson SeaBat 8125 multibeam echosounder (Teledyne, Denmark) mounted in the INNDAGA vessel (GEOMA-University of Vigo). This echosounder emits 240 beams, 40 times per second, with an angular resolution of 0.5° in the transversal direction and 1° in the advance direction. The obtained raw field data was corrected to remove errors of a different nature: (a) the GPS positioning error, corrected by the DGPS system (Differential GPS); (b) the error due to the vessel's movement (heave, pitch and roll), corrected by RTK (Real Time Kynetics) movement and a course sensor; (c) the error due to the tide-produced water level, corrected by pressure sensor field measurements; and (d) the error due to the sound velocity vaRiation produced by the changes in the water column density, corrected by SVP (Sound Velocity Profiler) measurements. The final resolution of the bathymetry was of 1 m in the horizontal plane and of a few centimetres in the vertical plane.

The detailed characterization of the superficial features of the present-day pipelines and the surrounding seabed up to a 30 m depth was performed with a Klein System 3900 (Klein Associates, USA) Side Scan Sonar (SSS) in December 2011. The GPS positioning during the acquisition was corrected with DGPS (Differential GPS). The acquisition and processing were carried out with SonarPro (Klein Associates) software.

Based on previous information of surface sedimentary distribution in the Ria de Vigo [19], the sediment samples were taken (in 16 points) following a regular grid of approximately 400 m using Van Veen grab from the INNDAGA vessel. The positioning was obtained with a GPS system.

3.2. Laboratory Work

The grain size of the sediment samples was obtained by dry sieving for the coarse fraction (>63 μm) every phi. The fine fraction (<63 μm) was separated by wet sieving, and was analyzed using a sedigraph (Micromeritics Sedigraph 5100, EEUU). The sediment grain size distribution was calculated using Gradistat v.7 software for Microsoft Excel [28]. This software analyses the statistically results of the granulometry using the method of moments, and classifies the samples by applying the graphic method [29,30]. The main parameter used in this work is the median (d50).

3.3. Database

The submerged relief of the Ria was characterized previously with bathymetric data of the nautical charts published by the Spanish Navy Hydrographical Institute (IHM). This data covers the Ria de Vigo above a 100 m isobath in the continental shelf with a highly variable spatial resolution between 10 and 200 m.

The wave characteristics in open waters were obtained from the Puertos del Estado database (http://www.puertos.es/es-es/oceanografia/Paginas/portus.aspx). The coastal buoy of Cabo Silleiro

(42.10° N, 8.93° W, Figure 1), located 3 km from the southern mouth of the Ria de Vigo, and the node no. 1044069 of the WANA numerical wave model 6 km from the northern mouth (42.25° N, 9.00° W, Figure 1) were selected. The buoy data supplied information about the seasonal, annual and extreme wave regime in terms of significant wave height and peak period, whilst the WANA numerical model supplied modelled data about the significant wave height, peak period and incoming direction of the different wave conditions.

The tidal data was obtained from the tidal gauge located in the harbour of Vigo (42.24° N, 8.73° W, Figure 1). The data used in this work was the water level, with a temporal resolution of one minute, and the harmonic constituents calculated from the historic time series of the gauge.

3.4. Numerical Simulations

Numerical modelling was used to determine the sediment transport under different wave conditions and bathymetry, which was modified in order to include the different route alternatives of the new pipeline. The numerical simulation of tidal currents, waves and sediment transport was performed using the open source version of Delft3D (Deltares, The Netherlands). The software consists of various modules: hydrodynamics (FLOW), waves (WAVE) and sediment transport (MOR). Delft3D computes the simulations by the finite differences method, and the configuration used in this study was in two horizontal dimensions (2DH).

In the hydrodynamic simulations, the program solves the movement equations of the flow, known as the momentum, continuity and conservative transport equations. The wave module (WAVE) is the Simulating Waves Nearshore (SWAN, Delft University of Technology, The Netherlands) [31]. This program solves the wave energy balance equation and takes into account the bottom friction, the current, the wind effect, the wave-wave interaction, the refraction and the reflection. The FLOW and WAVE modules share the information between them; in this way, the wave module takes into account the results of the flow module and vice versa. The morphodynamics module (MOR) uses the results from the previously mentioned modules to compute the sediment transport, both bed and suspended loads, updating in each time step the bottom sediment level and the bathymetry resulting from these processes. The calculations are based mainly on the critical shear stress of the sediment, which depends on the sediment median, d50. This data was obtained from the field campaign samples. The input parameters necessary to simulate the current and the water level were the main harmonic constituents of the tide available in the database of the tidal gauge of the Vigo harbor.

The model was established with a domain covering the entire Ria de Vigo, which is 22.8 km wide and 37.3 km long. A curvilinear grid was applied on this domain that was adapted to the geometry of the Ria, making the orientation of the cells coincide with the orientation of the south and north Ria mouths, respectively, and with the axis of the estuary. The grid cell size varied from 220 m in the open sea to 8 m at Samil Beach, depending on the Ria's geometry. The grid quality is critical for accurate simulations; therefore, the grid was made sufficiently dense in the area of interest, with orthogonal values of less than 6% in close boundaries and 2% in open boundaries.

The initial bathymetry was based on three different sources: a) the nautical chart of the Ria de Vigo of Instituto Hidrográfico de la Marina, with a resolution varying from 800 to 10 m, inversely proportional to the depth; b) a topographic survey on the emerging areas of Samil and Vao Beaches; and c) a bathymetric survey on the subtidal area of the Samil and Vao Beaches using a multibeam. Spatial resolution of these sources was higher than the resolution of the model grid, so a grid cell sea bottom was obtained by the "nearest point" option.

The model was 2DH, and had three open boundaries of two types: the western (open sea) boundary, located 8.5 km away from the Cíes Islands, at a depth of approximately 100 m, is a water level type, and the northern and southern boundaries are of the Neumann type. The water level type boundary calculates the tide by means of the known harmonic components M2 and S2 (main lunar half-day and main solar half-day, respectively). The Neumann-type boundary calculates the flow from water level gradients. This configuration is appropriate in the case that the model contains transverse boundaries to the coast with an open sea boundary, thus minimizing the generation of false waves within the model. Wave parameters are entered from all three boundaries simultaneously.

The input data for the wave modelling (significant wave height, peak period, and provenance direction) were grouped in different cases. In the selection of the cases, low energy (named summer) and high energy (named winter) conditions were taken into account. In addition, the sensitive incoming directions for the area of interest were also selected. The presence of the Cíes Islands in the entrance produces a shielding effect within the Rias with respect to the incident open sea waves. Therefore, the wave conditions most affecting the area of interest are those that coincide with the natural orientation of the Ria mouths; i.e., SW (225°) and NW (315°). The Hs-Tp correspond to the most frequent combination (low energy case) and to the most frequent considering high energy conditions (assuming high energy when Hs > 2 m and Tp > 14 s). The simulated cases are summarized in Table 1.

Table 1. Simulated wave cases, including the frequency of each incoming direction and the frequency of the Hs-Tp values in % of time.

Cases	Direction (°)	Hs (m)	Tp (s)	F. Direction %	F. Seasonal %
SW winter	225	2.5	14	7.5	2.5
SW summer	225	1.5	10	7.5	10.5
NW winter	315	2.5	14	50.6	2.5
NW summer	315	1.5	10	50.6	10.5

The model simulated 10 days for each wave condition, with a morphodynamic acceleration factor of 5; that is, the morphologic changes in each simulation were equivalent to 50 days. This time was enough to observe the sediment transport effects in the beach, as well as the interaction between the sediment dynamics and the planned structure under different wave conditions. The morphologic response of the beach to each wave condition was analyzed by transversal profiles obtained from the resulting bathymetry and erosion-sedimentation maps of the seabed. The model ran in a unique bathymetric grid, in which the resolution was locally increased to represent the submerged pipeline.

3.5. Birkermeier (1985) Equation of Closure Depth

The theoretical closure depth in the present work was calculated by the equation proposed in [32]. This equation relates the closure depth profile to the most energetic waves:

$$h^{*} = 1.75\, H_{12} - 57.9\left(\frac{H_{12}^{2}}{g\, T_{12}^{2}}\right) \tag{1}$$

where H_{12} is the significant wave height exceeding 12 h per year, T_{12} is the period linked to H_{12}, and g is the acceleration due to gravity. The data to determine H_{12} and T_{12} were obtained from ROM 0.3–91 [24]. The H_{12} was calculated from the mean significant wave height regime. The period linked to the wave height is defined in ROM 0.3–91 for Area III as

$$T_p = (4.5 \sim 9.2)\sqrt{H_s} \tag{2}$$

In the above equation, the peak period is proportional to the Hs by a coefficient that may vary within a range of values, highlighting that there is no lineal relationship between these two parameters.

4. Results

4.1. Bathymetry

The submerged area of the Samil Beach is characterized for combining sandy zones with rocky outcrops (Figure 2a). Above a 10 m isobath, the subtidal zone is bounded in the north by the Cabo de Mar rocky outcrop (a) and in the south by the outcrop dividing the Samil and Vao Beaches (b). In the central sector, there are also a group of outcrops (c, d and e) in which the two old submerged pipelines are located. The slope in the southern zone is less pronounced (0.027) than in the northern zone (0.036). Given an isobath of 10–20 m, the seabed is formed mainly by sediment, excepting a series of rocky outcrops (f) located at a 15 m depth. In this area, there is a group of marks in the seabed that could be anthropic, possibly tracks left by some fishing gear. The slope in this area is 0.009. This sector is bounded in the deepest zone by the escarpment (20 m) that limits the central channel of the Ria. In the escarpment, two sections with different slopes are observed—one much steeper (0.063) (h) than the other one (0.032) (g). In the latter section (g), 300 m long ripples with wavelengths of 15 m are observed. The SW zone of Samil Bay is characterized by the rocky outcrop that forms Toralla Island. In the shallowest sector (0–10 m), between Vao Beach and the island, a lobe-shaped double sand bar is developed (i).

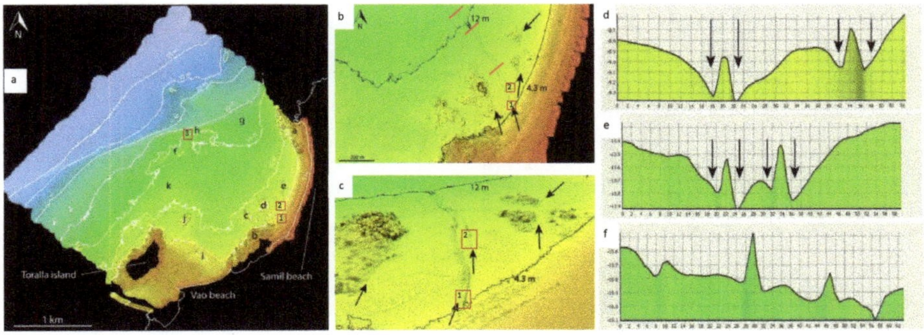

Figure 2. (**a**) Multibeam echosounder bathymetry with significant morphologic elements highlighted (in black lettering, see text). Lines in blue represent the route of ancient pipelines. (**b**) Plan view detail. Red lines define the position of the profiles shown in d, e and f. (**c**) Perspective of the multibeam bathymetry with 4.3 and 12 m isobaths. It is possible to appreciate the furrows in the surroundings of the rocky outcrops and the pipelines (tagged with arrows) and the accumulation of sediment in the space between them. Blue points mark the position of Side Scan Sonar (SSS) (**d**) Transversal bathymetric section of the pipelines, in the W–E (left–right, respectively) direction, showing the two pipelines at a 7 m depth. (**e**) Transversal bathymetric section in the W–E (left–right, respectively) direction, showing the two pipelines at a 12 m depth. (**f**) Transversal bathymetric section in the W–E (left–right, respectively) direction, showing the two pipelines at a 14 m depth. The scale of the axis is not the same. Erosion furrows are tagged with black arrows. Red squares represent the location of SSS in Figure 4.

In the bathymetry, the old pipelines can be easily distinguished in the seabed. Both pipelines are completely buried in the shallowest area (Figure 2a–c). The occidental pipeline (OCP) emerges at 4.3 m, whilst the oriental pipeline (ORP) remains buried up to a 6 m depth, the ORP arriving to a 25 m isobar and the OCP to a 30 m isobar, already in the central channel.

The resolution of the bathymetry allows one to distinguish some morphologic characteristics linked to the obstacles present at the bottom (rocky outcrops and current pipelines) of the shallowest zone (0–10 m) at Samil Beach (Figure 2b,c). The rocky outcrops present erosion structures 5–10 m long and 0.5 m deep on the SE side, and sediment accumulation in the NW side. In the pipelines, the OCP presents erosion at both sides up to a 12 m depth (Figure 2d–f) that is 4–6 m long and 0.5 m deep. The ORP presents a larger erosion furrow, up to 8 m, with a similar depth, but in the east flank only. This furrow also disappears at a 12 m depth (Figure 2d–f). The magnitude of these furrows parallel to the pipelines decreases gradually as the depth increases. The space between the two pipelines is characterized by an asymmetric accumulation that is developed in the W–E direction (Figure 2f).

4.2. *Surficial Grain Size Distribution*

The subtidal sector of Samil Bay is dominated by sands (Figure 3). The superficial distribution presents a textural transition in the SW–NE direction from very coarse sand (>1000 µm) to very fine sand (200–100 µm). In the NE corner of the studied area, the presence of mud was detected in low percentages (<10%).

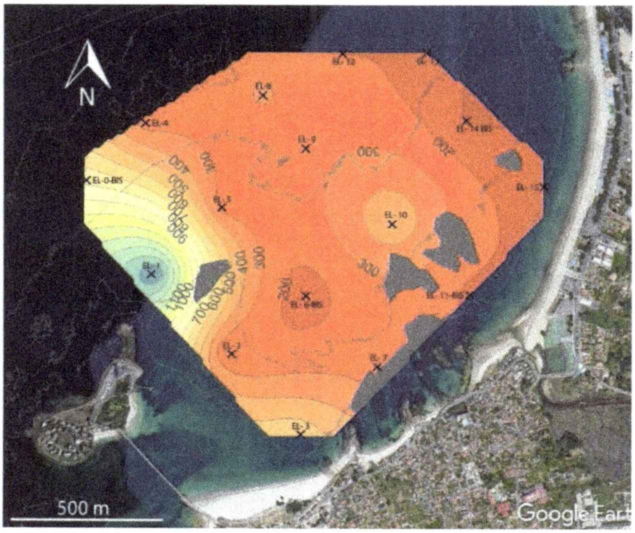

Figure 3. The median grain size distribution of the sand fraction in the study area, calculated using kriging interpolation with the default linear variogram. Maximum values are in blue (1400 µm), and minimum values are in red (100 µm). Sample positions are marked with an X. Grey areas correspond to rocky outcrops. The isobaths from 0 to 35 m are also shown.

Toralla Island and the submerged outcrops exert important hydrodynamic control on the sediment distribution in the bay. These geographical features produce a shadowing effect, favouring the finest material accumulation on their lee side (<200 µm).

4.3. *Side Scan Sonar (SSS)*

The bottom characterization by the side scan sonar allowed for an assessment of the current state of the old pipelines and the evaluation of the interaction between the seabed and the structures, in addition to gathering information about the sedimentological characteristics of the seabed.

The shallowest images, at 3.3–7 m deep (Figure 4a), showed the emerging zone of the occidental pipeline (OCP). It can be observed that there is an accumulation in the west flank, at some point burying the pipeline, and erosion in the east flank with parallel accumulation, as was identified in the multibeam echosounder dataset. In some images, this pipeline shows an unearthed concrete basement. In the images at a deeper depth, between 7 and 12 m, the oriental pipeline (ORP) is identified (Figure 4b), and a similar accumulation-erosion pattern in the surroundings of both structures is observed, but of a smaller magnitude than in the shallowest section. At isobaths at least 12 m deep, the interaction between the pipelines and the seabed disappeared (Figure 4c), and the images at a deeper depth did not show interaction marks anymore.

The main sedimentary characteristics of the seabed in this area were well-developed ripple fields due to the wave action. One of those fields was identified 300 m eastward from the pipelines, at an

11 m depth. Another field was observed in the vicinity of the pipelines at a depth between 15.1 and 15.3 m, occupying a 0.41 km^2 area with a maximum height of 0.2 m.

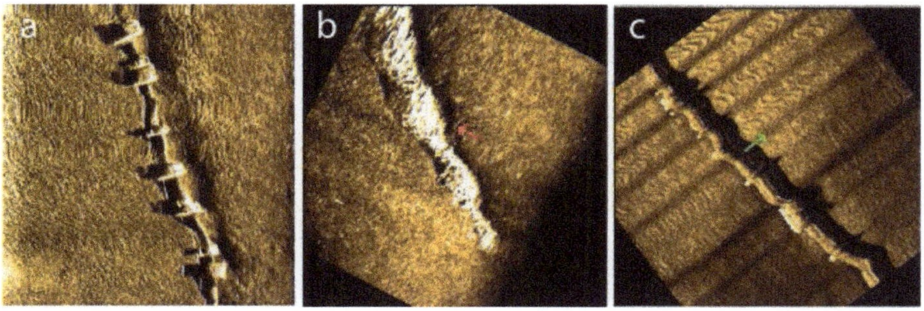

Figure 4. SSS images: (**a**) the occidental pipeline at a 4.3 m depth; (**b**) the oriental pipeline at a 5.2 m depth protected by a concrete cover; and (**c**) the occidental pipeline at a 15.1 m depth in a ripple field. The images are oriented to the north. The location of these sonograms is in are marked as red squares in Figure 2.

4.4. Seasonal Dynamics of Samil Beach

The seasonal dynamics of Samil Beach are controlled by the wave action. The wave energy in this area is conditioned by the amplitude and orientation of the Ria mouths, as well as the beach location and orientation. From the wave database, we identified two main incoming directions that directly affect the area of interest. One is the NW direction, the most frequent direction in this area, which coincides with the natural orientation of the northern mouth. This allows these waves to propagate southward inside the Ria and towards Samil Beach, which is exposed because it faces the W-NW direction. The second main incoming direction is SW, less frequent than the previous case, but linked to storms, which coincides with the orientation of the central channel of the Ria. Such waves reach Samil Bay after a refraction processes due to the southern margin coastline in the outer Ria sector and the presence of the Toralla rocky outcrop.

Under NW winter conditions (315°, Hs_0 = 2.5 m) in the outer sector of the Ria, the significant wave height (Hs) diminishes rapidly until it reaches 1 m (Table 2). The waves arrive in the area of interest with a 0.5–0.8 m height, presenting a propagation coefficient of 20–32%. The incoming direction changes from 315°, in the open sea, to approximately 310° in the area of interest. Under similar conditions, but in the SW incoming direction (225°), the waves are attenuated as they travel inside the Ria through the central channel, reaching values of 1–1.5 m in the outer sector of the Ria. In the area of interest, they present 0.8–1 m, showing a propagation coefficient of 32–40%. The change in the incoming direction is greater under these conditions due to the refraction along the southern margin—from 225° in the open sea to 300° in the area of interest.

Under summer conditions (Hs_0 = 1.5 m), the Hs values inside the bay are much smaller, although the distribution pattern inside the Ria and the propagation coefficients are similar. Under NW conditions, the values in the bay are 0.3–0.5 m, showing propagation coefficients of 20–33%. Under SW conditions, in the area of interest, Hs values of 0.4–0.6 m are observed, presenting a propagation coefficient of 26–40%.

The study of the morphological response of the beaches to different wave conditions (Table 1) was based on the Delft3D model's (MOR module) results. The erosion-accretion results under winter conditions (Figure 5) showed that the sediment transport was produced exclusively along a narrow strip 300 m wide in the subtidal sector. In the two cases studied, the main feature was the transport produced in strips parallel to the coastline, accretion in the intertidal and shallow subtidal sector, and erosion in the deepest subtidal sector.

Table 2. Wave condition variation during the wave propagation inside the Ria.

Open Sea		Studied Area		Propagation Coefficient
H (m)	Direction	H (m)	Direction	
		Winter Conditions		
2.5	315° (NW)	0.5–0.8	310°	20–32%
2.5	225° (SW)	0.8–1.0	300°	32–40%
		Summer Conditions		
1.5	315° (NW)	0.3–0.5		20–33%
1.5	225° (SW)	0.4–0.6		26–40%

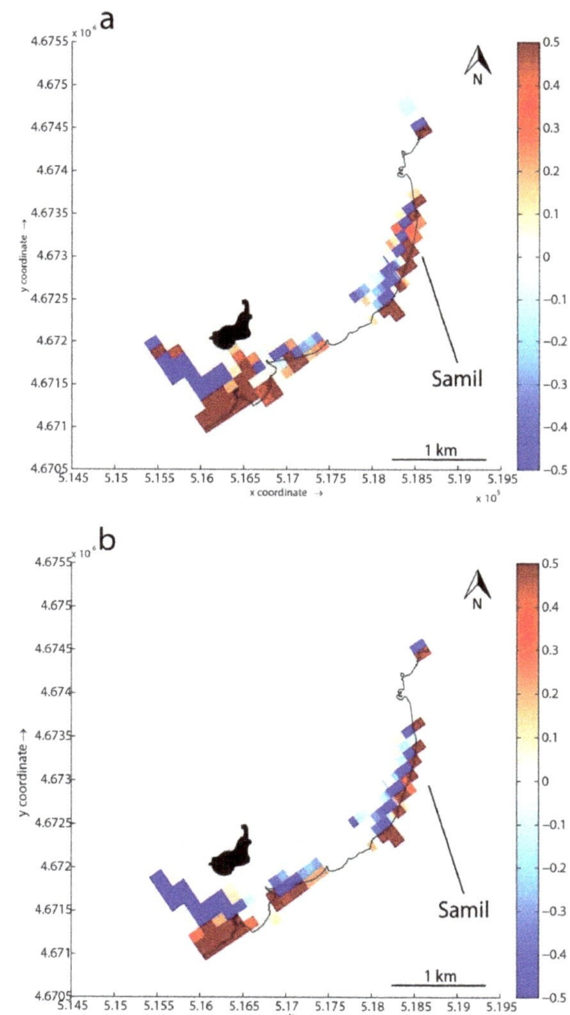

Figure 5. Modelling results showing the present-day erosion and accretion (meters) in the study area under NW (**a**) and SW (**b**) winter conditions (Hs = 2.5 m, Tp = 14 s).

The resulting sedimentation presented values of 0.3–0.5 m, and the erosion presented values of 0.25–0.5 m. Under summer conditions (not shown), the transport pattern was similar, but of substantially smaller magnitude, with an accretion of 0.3–0.35 m, and with an erosion of 0.15–0.35. The location of the accretion and erosion areas, and their magnitude change, was between the NW (Figure 5a) and SW (Figure 5b) wave conditions. The location varies specifically in the deepest subtidal sector, where accretion is produced under SW conditions, and erosion is produced under NW conditions. Based on the bathymetric results of the model, the seasonal changes from the transversal profiles were analyzed; four profiles in Samil Beach, one profile located in the transition zone between the two beaches, and one more profile in Vao Beach (Figure 1c).

In general, the resulting profiles in the sandy area (P1, P2 and P3, see Figure 1), under winter and summer wave conditions, present a higher slope than the original profile (Figure 6). This favours the existence of a cutting point between the original and resulting profile. Thus, in shallower depths than this point, it produces accretion, whilst at deeper depths, it causes erosion in the profile. Comparing the summer and winter conditions, the summer profiles have higher slopes and, consequently, cut the original profile at a shallower depth (between −1 and −2 m) than the winter profiles (between −2 and −4 m). In summer conditions, for both NW and SW waves, the beach response was similar. In winter conditions, the profile response varies slightly, developing steeper slopes with NW waves. The slope of the profiles shows an increasing trend towards the south. In P1 (Figure 6a), the resulting slope varied slightly between 0.040 and 0.047; in P2 (Figure 6b), it varied between 0.038 and 0.070 in winter and between 0.026 and 0.068 in summer; in P3 (Figure 6c), it varied between 0.059 and 0.070 in winter and was 0.050 in summer.

The sedimentation, as well as the erosion, of the original profile also tends to increase towards the south, from P1 to P3. Thus, the accretion values varied between 0.3 and 0.8 m in P1, between 0.6 and 1 m and between 0.7 and 1.6 m in P2, and they were 1.3 m in P3. Regarding the erosion, the values varied between 0.5 and 1.0 m in P1 and P2, and between 0.25 and 2.1 m in P3. Under summer conditions, more sedimentation was observed in the intertidal sector of P1 (0.6–0.7 m) and P2 (0.6–0.75 m), whilst this is linked to winter conditions in P3 (1.0–1.25 m). The highest erosion in the subtidal sector is associated with winter conditions in the three profiles (0.65–2.1 m). The analyses of seasonal changes in the Samil Beach profile allowed us to determine the deepest point at which sediment movement is observed. The maximum movement depth is −7.1 m in P1 and P3 (Figure 6a,c) and is −7.4 m in P2 (Figure 6b).

The profile P4, located in an area of rocky outcrops (Figure 6d), presents erosion over the entire profile, with values of 0.75–1 m under winter conditions and a value of 0.5 m under summer conditions. The maximum sediment movement depth in this profile is located at a −6 m isobath. In all cases, the final slope is similar to the original one (0.033).

The profiles P5 and P6 (Figure 6e,f) are located in a zone between Samil and Vao Beaches, and in Vao Beach, respectively. The seabed of P5 consists of rocks up to a 5 m depth. The profile P6 presents a similar behaviour to P3, showing a cutting point between the original and resulting profiles that defines the accretion zones at shallow depths and the erosion zone in the deepest part of the profile. The accretion is 0.75 m, whilst the erosion is 0.25 m. The slope of the resulting profile in winter conditions is 0.032, higher than the original, which is 0.026. In this case, the maximum sediment movement observed is at a −4.5 m depth.

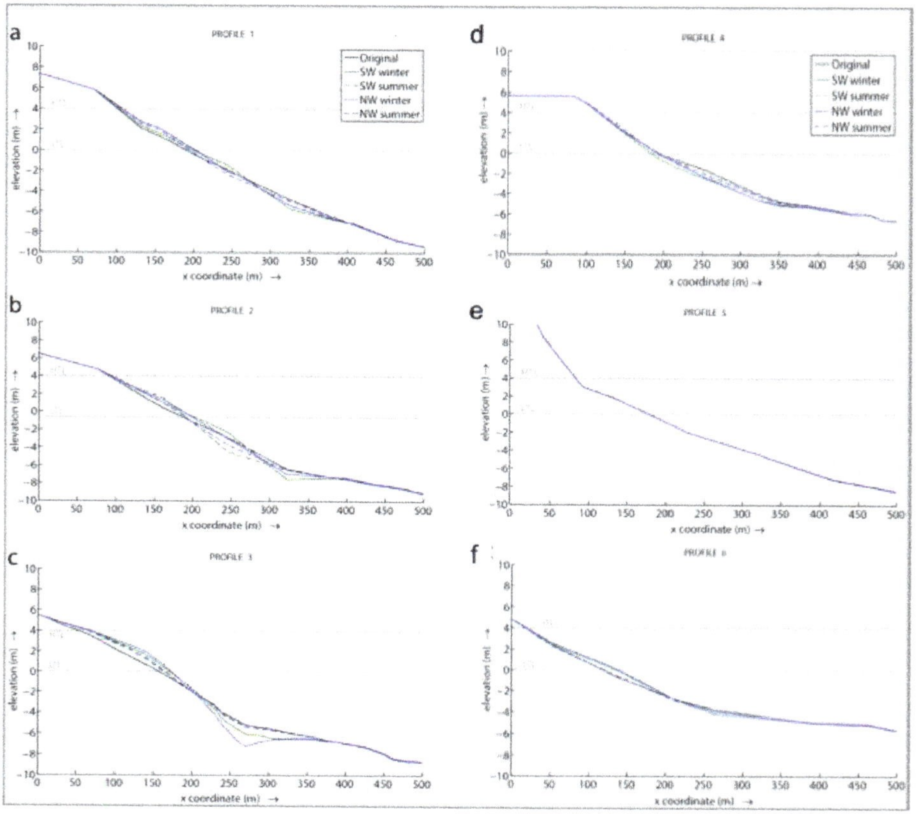

Figure 6. Topobathymetric profile evolution under different wave conditions. The x-axis presents the distance in meters from the profile head, and the y-axis represents the elevation (negative) in meters. High tide maximum limit (HTL) and low water minimum limit (LWL) are highlighted in gray. The x- and y-axes have different scales. Letters a to f refer to the individual beach profiles locations indicated in Figure 1.

4.5. *Modelling of the Different Route Alternatives*

The simulation of the mutual interaction between the structure and natural sediment dynamics in the beach was performed for the two possible pipeline route alternatives. For each alternative, the previously described four wave cases were simulated; NW summer and winter and SW summer and winter. The results described hereafter refer to winter (storm) cases, since in the summer cases the presence of the new pipelines did not cause any significant morphologic alteration.

4.5.1. Southern Route Alternative

Under SW storm waves (not shown), the area of the bay where the pipeline construction is planned is partially sheltered from the wave action by the presence of Toralla Island. Thus, the waves reach the southern corner of the bay with an incoming direction of 300°, practically parallel to the pipeline route. The Hs showed locally higher values (≈0.65 m) over the pipeline, between 5 m and 10 m deep, than in the surrounding area (≈0.55 m). This local increase was also reflected in the orbital velocity, with 0.30 m/s over 0.27 m/s. Along the pipeline route, sediment movement is detected eastward from the shallowest section. The variation in the wave conditions in the shallowest section of the structure produced an erosion of 0.20 m in its east flank with respect to the results without a pipeline presence.

This interaction disappears as the depth increases along its route. The results of the model reflected that the structure was located in a low mobility area, since the shallowest sector of the beach, located above the pipeline emerging point, consisted of a rocky seabed. Under these wave conditions, the presence of the structure did not produce any effect in the natural dynamics of the beach (Figure 5a).

Under NW storm wave conditions, the southern corner of the bay is exposed to the incident waves, and Toralla Island does not have a strong influence, as in the previous case. The Hs in the southern corner of the bay is between 0.5 and 1 m, again in a similar incoming direction to the pipeline route (300°). Similarly, with respect to the previous case, in the shallowest section of the pipeline (<8 m), a variation in the Hs over the structure is produced −0.6 m over it and 0.55 m in the surrounding area. This increase was also produced in the bottom orbital velocity −0.25 m/s over 0.22 m/s. Thus, the alterations produced in the waves only cause sediment transport in the shallowest section of the pipeline. The presence of the pipeline generates a flank erosion furrow that is similar to the previous case but of smaller magnitude, 0.11 m, in the east flank, with respect to the results without a pipeline. The analysis of the erosion/sedimentation processes in Samil Beach revealed that the presence of the pipeline did not alter the natural behaviour of the beach. Comparing the erosion/sedimentation results with the simulation with no structure (Figure 5b), it is also observed that the pipeline did not produce a shielding effect in the sediment transport in this case, the erosion/sedimentation zones and values being similar.

Ultimately, the southern route alternative showed erosion on the east flank in the shallowest sector of the pipeline related to the alteration of the waves due to the pipeline's presence. However, this alteration was local and did not affect the global sediment dynamics of the beach.

4.5.2. Northern Route Alternative

The northern corner of the bay is more exposed to the waves than the southern corner. However, in key terms, the presence of the structure did not affect the waves in the bay. Under SW storm waves, a local increase in the Hs in the shallowest section was also detected (Figure 7a), with values of 0.7 m over 0.6 m, equally reflected in the bottom orbital velocity (Figure 7b) −0.35 m/s over 0.29 m/s. The incoming direction of the waves that reach the beach is 310°, an angle of 10° with respect to the pipeline. The shallowest section of the pipeline (<8 m) is located in an area of general erosion (Figure 7c). The presence of the pipeline in this sector produced an alteration of 0.15 m in its east flank followed by a 0.11 m accretion with respect to the results without the structure (Figure 7d), whilst at deeper depths, no interactions were observed.

The presence of the planned pipeline under SW storm waves did not alter the general transport pattern of the beach (Figure 7c) with respect to the observed pattern without a structure (Figure 5a). The areas where sedimentation/erosion was produced were the same, with a similar range of values, and no shielding effect in the sediment pattern was detected.

The sediment dynamics under NW winter storm waves changes with respect to the previous case (SW). The waves' incoming direction in this case coincides with the pipeline's orientation. The values of the wave parameters, Hs, and the bottom orbital velocity are significantly smaller than they are under SW winter storm waves. These parameters increase over the structure in the shallowest section (<8 m), showing a difference of 0.1 m in the Hs and 0.05 m/s in the bottom orbital velocity, with respect to the surroundings (Figure 8a,b, respectively). The shallowest segment is located in an erosion area, similar with respect to the previous modelled case (Figure 8c) but of smaller magnitude (Figure 8d). In this case, the presence of the pipeline produced minor erosion at its flanks (0.07 m) due to the lower wave energy.

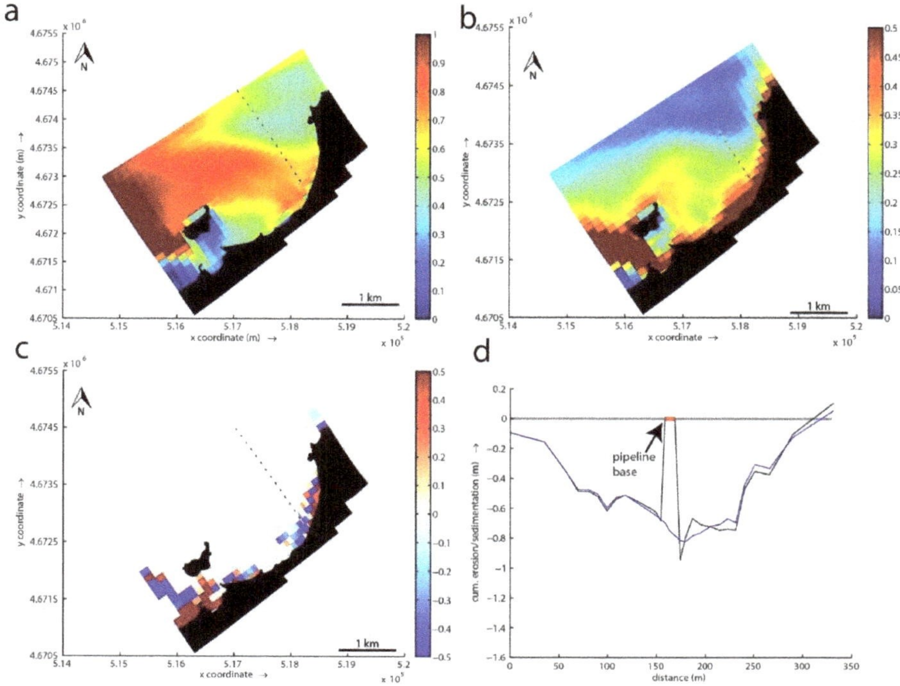

Figure 7. Northern alternative case under SW wave storm in the area of interest: (**a**) significant wave height; (**b**) bottom orbital velocity; (**c**) accumulated accretion and erosion in the area of interest (the pipeline route is highlighted with black points); (**d**) transversal view showing the accretion and erosion with (black) and without (blue) the pipeline at a 6 m depth (the pipeline base is highlighted with a red line).

The natural beach dynamics in this case were not altered by the presence of the planned pipeline either, since the waves were barely altered. Comparing the resulting erosion/sedimentation with the results without a pipeline, the sediment transport was the same in both cases.

According to the modelled transport results of southern and northern route alternatives under different storm conditions (SW, NW), two areas can be clearly distinguished in the beach—the shallow zone (between 0 and 8 m deep) and the deep zone (more than 8 m deep). The shallow zone is where the main seasonal sediment movement occurs, producing erosion-sedimentation strips parallel to the shoreline, especially in the norther corner. In the deep area, no sediment movement was detected. The presence of the pipeline in the shallow zone produces a local interaction with the sediment transport, causing erosion zones at both pipeline flanks. Nevertheless, the presence of the pipeline does not alter the global sediment transport pattern in the beach.

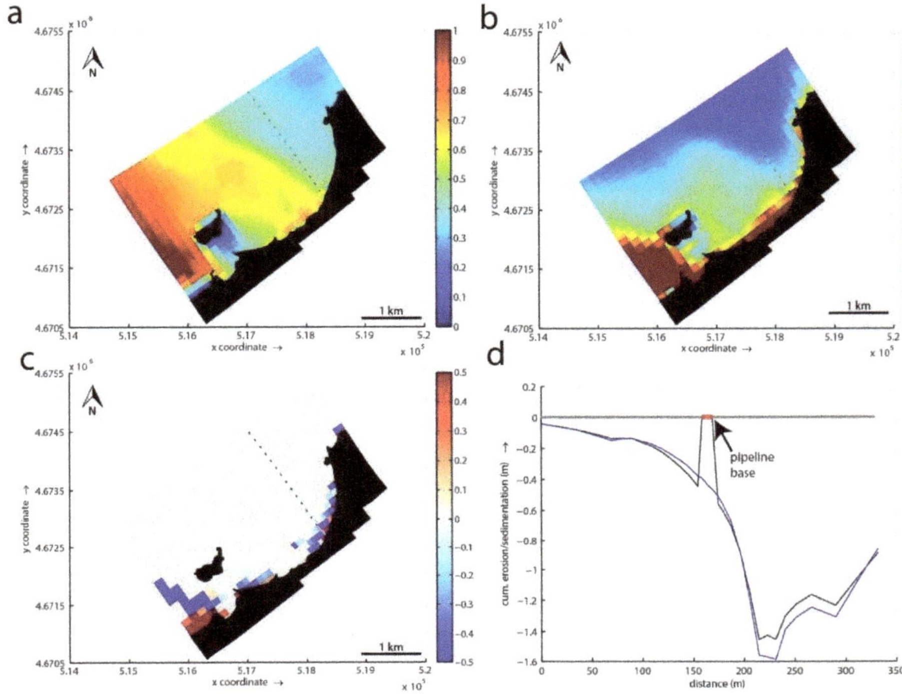

Figure 8. Northern alternative case under NW waves storm in the area of interest: (**a**) significant wave height; (**b**) bottom orbital velocity; (**c**) accumulated accretion and erosion in the area of interest (the pipeline route is highlighted with black points); (**d**) transversal view showing the accretion and erosion with (black) and without (blue) the pipeline at a 6 m depth (the pipeline base is highlighted with a red line).

5. Discussion

When a hard intervention is performed in a highly dynamic sedimentary system, as in a beach, it must fulfil security, service and exploitation requisites. In this work, the security of the pipeline and Samil Beach was analyzed, allowing for the selection of the most feasible route alternative and identifying future interactions that would occur between the planned route alternatives and the sediment beach dynamics.

5.1. *Effect of the Pipeline Route Alternatives on the Beach and Alternative Selection*

All data shows that Samil Bay is a low energy zone, in which the waves that arrive are very attenuated with respect to the open sea conditions, between 60% and 75%, due to the Ria de Vigo's geomorphology. In addition, the presence of Toralla Island exerts a local shielding effect in the southern part of Samil Bay. Thus, the most exposed part of the bay to the wave action coincides with depths of more than 20 m. In the same way, the resulting energy due to the wave shallowing process is concentrated in a narrow strip along the coastline at depths less than 8 m. In addition, the beach sediment shows that the beach is a low energy area where the prevailing mean grain size is lower than 300 μm (medium, fine and very fine sands).

The numerical model Delft3D has been shown to be a useful tool for the assessment of submerged pipeline construction alternatives. The results of the numerical model showed that the pipelines of the two alternatives did not alter the global wave pattern within the bay. Regarding the sediment movement, it was determined that the two suggested routes produced a low affection in the subtidal area of Samil Beach. Only in the shallowest (<8 m) sector of the beach did the presence of the pipelines cause small-scale alterations at the bottom. The field data (bathymetry and SSS) supported the model results, determining that the highest interactions are produced locally at depths lower than 12 m, diminishing as depth increases. Accordingly, the presence of rocky outcrops, some of them large, seems to produce a limited effect on the sediment dynamics since the bathymetry does not reveal strong interaction signs. In the assessment of the best-suited pipeline routes, taking into account the structure security criteria, the results showed that, in the two planned route alternatives, the interaction was similar. Therefore, this criterion was not crucial in the route selection. Nevertheless, the different characteristics of the bottom between the northern and southern routes conditioned constructive aspects at each alternative and, consequently, the work cost. Thus, the alternative selection was based mainly on economic criteria, and the northern route was finally selected.

5.2. Pipeline Emerging Depth in the Northern Route

The constructive design of the northern route alternative consisted of a 620 m long, buried section from the water treatment plant to Samil Beach, and of a second, 3040 m long, submerged section that emerged gradually in the subtidal area of the beach, lying on the seabed for the rest of the route.

For this reason, the definition of the pipeline's emerging depth was a critical constructive point in the project. The selected depth must also fulfil the security criteria of the structure. The beaches are environments of high temporal variability, where small changes in the wave conditions may lead to important changes in the sediment dynamics and in the morphology of the emerged and submerged sectors. Thus, the maximum sediment movement depth is a crucial parameter in the pipeline design. In engineering terms, this depth is known as the closure depth, a theoretical concept that establishes the depth from which we assume there are no significant morphological changes in the bottom. The definition of this depth allows for the establishment of a security depth for the structure placement.

The closure depth was addressed by a multitool approach in order to increase the reliability of the solution. Thus, existent scientific literature data, theoretical calculations by the Birkemeier equation [32], fieldwork data (multibeam echosounder bathymetry and SSS) and numerical modelling results were compared. All of them were calculated to a zero sea level reference of the Vigo harbor.

Previous studies in Samil Beach [33] established that the closure depth was located between 8 and 12 m in low energy conditions, and up to 17 m in high energy conditions.

The theoretical closure depth in the present work was calculated by Equation (1). For the relation between H_{s12} and T_p (Equation (2)), a range of values was obtained in the closure depth calculations; a minimum value was associated with the lowest peak period ($T_{12,\,min}$), $h^*_{min} = 10.36$ m, and a maximum value was associated with the highest period ($T_{12,\,max}$), $h^*_{max} = 12.47$ m. A range of values were obtained in the closure depth calculations; a minimum value associated to the lowest peak period ($T_{12,\,min}$), $h^*_{min} = 10.36$ m, and a maximum value associated to the highest period ($T_{12,\,max}$), $h^*_{max} = 12.47$ m. The study of the bathymetry and the SSS images highlighted the existence of erosion structures at both flanks of the old pipelines. These structures were identified up to approximately a 12 m depth. This can be considered to be the maximum depth at which the sediment can be moved due to dynamical processes. Therefore, this can be interpreted as direct data regarding the beach closure depth.

The use of the Delft3D numerical modelling and the analysis of transversal beach profiles and its temporal evolution established a maximum profile variation at around a 7 m depth. From this point on, there were no changes in the morphology. Though the model has not been validated with measurements, comparison with in situ data acquired by other technical means shows that the model slightly underestimates the morphodynamic beach behavior.

In Table 3, depth values determined from different tools and sources are summarized:

Table 3. Closure depth determined from different tools and sources.

Data Type	Closure Depth (m)
Scientific literature	8–12
Birkemeier (1985) equation	11.9 (10.36–12.47)
Multibeam echosounder bathymetry	12
Side Scan Sonar	12
Delft3D model	7.4

The closure depth data determined from the visual inspection of the bathymetry and SSS images coincides with the main terms of the Birkemeier formula [32], even though this was developed for open sea beaches. This is most likely due to the extreme storm events' effect on the obstacles. When these short but very energetic events take place, these waves interact with the obstacles located at deeper depths than those that usually produce erosion/sedimentation structures. These structures would have a permanent character, since the usual wave conditions could not modify them due to the depth.

The closure depth given by the scientific literature and numerical modelling is smaller than the theoretical and observed one. The scientific literature presents a range of values whose upper limit (12 m) coincides with the closure depth determined by the Birkemeier equation [32]. The numerical modelling establishes the seasonal variability of the beach profile up to a 7.4 m depth, with a horizontal forward-retreat of the profile that can exceed 100 m. This model is not calibrated in absolute terms, but it is important to highlight that the resulting data is within the range of the beach morphodynamics under a similar hydrodynamic regime.

6. Conclusions

Several data collection techniques and numerical modeling approaches have been integrated to explore the interaction between two pipeline construction alternatives and Samil Beach. The submerged pipeline planned for Samil Beach would not affect the global sediment dynamics of the beach, based on the simulated cases. Most of the pipeline is outside the sector where the main transport processes take place, avoiding a shielding effect in the main transport paths. Thus, the seasonal morphodynamic behaviour of the beach would not be affected.

Nevertheless, the structure would be affected by the beach sediment dynamics, producing mainly erosion in its surrounding seabed until the closure depth is reached. This interaction does not hazard the integrity of the pipeline in the case that it emerges deeper than the closure depth. Taking into account the values obtained for the closure depth in the beach (8–12 m), the pipeline emerging point must be located above an 8 m deep isobath in order to avoid its integrity hazard, which represents the seasonal beach sediment movement. The results also indicate that, if the pipeline emerges on a seabed above a 12 m bath isobath, its integrity would be guaranteed.

Thus, the pipeline-beach interaction criterion for the best suited pipeline alternative selection has been discarded. The multitool strategy has proved to be efficient in assessing the sediment dynamics of the Samil Beach environment and the effects of each pipeline route alternative on the beach. It has also allowed us to identify the consequences of the seasonal sediment dynamics on the pipeline alternatives; something essential for the security of the structures.

This study highlights the advantage of exploring the effect of submerged infrastructures with a multidata approach to fully understand the diversity of potential influences on local beach behavior.

Author Contributions: Conceptualization, R.D., D.R., and A.M.B; methodology, A.M.B. and D.R.; software, A.L.-K.; formal analysis; A.L.-K.; writing—original draft preparation, A.L.-K.; writing—review and editing, A.M.B. and D.R.; supervision, D.R. and A.M.B.; project administration: D.R., R.D., and A.M.B; funding acquisition, R.D. and A.M.B. All authors have read and agreed to the published version of the manuscript.

Funding: This research was partially funded by Aguas de las Cuencas de España S.A. (Acuaes) and Spanish project FAREWELP CGL2015-66681-R.

Conflicts of Interest: The authors declare that there is no conflict of interest.

References

1. Brown, A.; McLachlan, A. Sandy shore ecosystems and the threats facing them: Some predictions for the year 2025. *Environ. Conserv.* **2002**, *29*, 62–77. [CrossRef]
2. Merkens, J.-L.; Reimann, L.; Hinkel, J.; Vafeidis, A.T. Gridded population projections for the coastal zone under the Shared Socioeconomic Pathways. *Glob. Planet. Chang.* **2016**, *145*, 57–66. [CrossRef]
3. Gracia, A.; Rangel-Buitrago, N.; Oakley, J.A.; Williams, A.T. Use of ecosystems in coastal erosion management. *Ocean Coast. Manag.* **2018**, *156*, 277–289. [CrossRef]
4. Rizzeto, F. Effects of climate change on the morphological stability of Mediterranean coasts: Consequences for Tourism. In *Climate Change Management*; Springer: Basel, Switzerland, 2020; pp. 761–775.
5. Marchand, M.; Sanchez-Arcilla, A.; Ferreira, M.; Gault, J.; Jiménez, J.A.; Markovice, M.; Mulder, J.; VanRijn, L.; Stănicăf, A.; Suliszg, W.; et al. Concepts and science for coastal erosion management—An introduction to the Conscience framework. *Ocean Coast. Manag.* **2011**, *54*, 859–866. [CrossRef]
6. Thorne, K.M.; Elliot-Fisk, D.L.; Freeman, C.M.; Bui, T.D.; Powelson, K.W.; Janousek, C.N.; Buffington, K.J.; Takekawa, J.Y. Are coastal managers ready for climate hange? A case study from estuaries along the Pacific coast of the United States. *Ocean Coast. Manag.* **2017**, *143*, 38–50. [CrossRef]
7. Siders, A.; Keenan, J.M. Variables shaping coastal adaptation decisions to armor, nourish, and retreat in North Carolina. *Ocean Coast. Manag.* **2020**, *183*, 105023. [CrossRef]
8. Cooper, J.A.G.; O'Connor, M.C.; McIvor, S. Coastal defences versus coastal ecosystems: A regional appraisal. *Mar. Policy* **2020**, *111*, 102332. [CrossRef]
9. Ranasinghe, R.; Turner, I.L. Shoreline response to submerged structures: A review. *Coast. Eng.* **2006**, *53*, 65–79. [CrossRef]
10. Schoonees, T.; Gijón Mancheño, A.; Scheres, B.; Bouma, T.J.; Silva, R.; Schlurmann, T.; Schüttrumpf, H. Hard Structures for Coastal Protection, Towards Greener Designs. *Estuaries Coasts* **2019**, *42*, 1709–1729. [CrossRef]
11. Molina, R.; Anfuso, G.; Manno, G.; Gracia, J. The Mediterranean Coast of Andalusia (Spain): Medium-Term Evolution and Impacts of Coastal Structures. *Sustainability* **2019**, *11*, 3539. [CrossRef]
12. Nemes, D.D.; Criado-Sudau, F.F.; Gallo, N.M. Beach Morphodynamic Response to a Submerged Reef. *Water* **2019**, *11*, 340. [CrossRef]
13. Torres-Freyermuth, A.; Medellín, G.; Mendoza, E.T.; Ojeda, E.; Salles, P. Morphodynamic Response to Low-Crested Detached. Breakwaters on a Sea Breeze-Dominated Coast. *Water* **2019**, *11*, 635. [CrossRef]
14. Ming, D.; Chiew, Y.-M. Shoreline changes behind detached breakwater. *J. Waterw. Port Coast. Ocean Eng.* **2000**, *126*, 63–70. [CrossRef]
15. Ruiz-Martínez, G.; Mariño-Tapia, I.; Mendoza Baldwin, E.G.; Silva, R.; Enríquez Ortiz, C.E. Identifying Coastal Defence Schemes through Morphodynamic Numerical Simulations along the Northern Coast of Yucatan, Mexico. *J. Coast. Res.* **2016**, *32*, 651–669.
16. Bouvier, C.; Castelle, B.; Balouin, Y. Modeling the Impact of the Implementation of a Submerged Structure on Surf Zone Sandbar Dynamics. *J. Mar. Sci. Eng.* **2019**, *7*, 117. [CrossRef]
17. Klonaris, G.T.; Metallinos, A.S.; Memos, C.D.; Galani, K.A. Experimental and numerical investigation of bed morphology in the lee of porous submerged breakwaters. *Coast. Eng.* **2020**, *155*, 103591. [CrossRef]
18. Bernabeu, A.M.; Lersundi-Kanpistegi, A.V.; Vilas, F. Gradation from oceanic to estuarine beaches in a ria environment: A case study in the Ria de Vigo. *Estuar. Coast. Shelf Sci.* **2012**, *102–103*, 60–69. [CrossRef]
19. Vilas, F.; Bernabeu, A.M.; Méndez, G. Sediment distribution pattern in the Rias Baixas (NW Spain): Main facies and hydrodynamic implications. *J. Mar. Syst.* **2005**, *54*, 261–276. [CrossRef]
20. Castro, M.; Gómez-Gesteira, M.; Prego, R.; Taboada, J.J.; Montero, P.; Herbello, P.; Pérez-Villar, V. Wind and tidal influence on water circulation in a Galician ria (NW Spain). *Estuar. Coast. Shelf Sci.* **2000**, *51*, 161–176. [CrossRef]
21. Huthnance, J.M.; Van Aken, H.M.; White, M.; Barton, E.D.; Le Cann, B.; Coelho, E.F.; Fanjul, E.A.; Miller, P.; Vitorino, J. Ocean margin exchange—Water flux estimates. *J. Mar. Syst.* **2002**, *32*, 107–137. [CrossRef]

22. Ruiz-Villarreal, M.; Montero, P.; Taboada, J.J.; Prego, R.; Leitao, P.C.; Pérez-Villar, V. Hydrodynamic model study of the Ria de Pontevedra under estuarine conditions. *Estuar. Coast. Shelf Sci.* **2002**, *54*, 101–113. [CrossRef]
23. Pérez-Arlucea, M.; Mendez, G.; Clemente, F.; Nombela, M.; Rubio, B.; Filgueira, M. Hydrology, sediment yield, erosion and sedimentation rates, in the estuarine environment of the Ria de Vigo, Galicia, Spain. *J. Mar. Syst.* **2005**, *54*, 209–226. [CrossRef]
24. Del Estado, P. *Recomendaciones para Obras Marítimas (ROM) 0.3–91, 1991: Oleaje*; Anejo, I., Ed.; Clima marítimo en el litoral español. Puertos del Estado; Ministerio de Transportes, Movilidad y Agenda Urbana: Madrid, Spain, 1991.
25. Rey, D.; Mohamed, K.J.; Bernabeu, A.; Rubio, B.; Vilas, F. Early diagenesis of magnetic minerals in marine transitional environments: Geochemical signatures of hydrodynamic forcing. *Mar. Geol.* **2005**, *215*, 215–236. [CrossRef]
26. Varela, R.A.; Rosón, G. A general study of the Spanish North Atlantic boundaries: An interdisciplinary approach. *J. Mar. Syst.* **2005**, *54*, 1. [CrossRef]
27. Lorente, P.; Sotillo, M.G.; Aouf, L.; Amo-Baladrón, A.; Barrera, E.; Dalphinet, A.; Toledano, S.; Rainaud, R.; De Alfonso, M.; Piedracoba, S.; et al. Extreme wave height events in NW Spain: A combined multi-sensor and model approach. *Remote Sens.* **2018**, *10*, 1. [CrossRef]
28. Blott, S.J.; Pye, K. GRADISTAT: A grain size distribution and statistics package for the analysis of unconsolidated sediments. *Earth Surf. Proc. Landf.* **2001**, *26*, 1237–1248. [CrossRef]
29. Folk, R.L. The distribution between grain size and mineral composition in sedimentary rock nomenclature. *J. Geol.* **1954**, *62*, 344–359. [CrossRef]
30. Folk, R.L.; Ward, W.C. Brazos River bar: A study in the significance of grain size parameters. *J. Sedimentol. Petrol.* **1957**, *27*, 3–26. [CrossRef]
31. Booij, N.; Ris, R.C.; Holthuijsen, L.H. A third-generation wave model for coastal regions 1. Model description and validation. *J. Geophys. Res.* **1999**, *104*, 7649–7666. [CrossRef]
32. Birkemeier, W. Field Data on Seaward Limit of Profile Change. *J. Waterw. Port Coast. Ocean Eng.* **1985**, *111*, 598–602. [CrossRef]
33. Vila-Concejo, A.; Alejo, I.; Vilas, F. Monitoring of a beach with a strong anthropogenic impact by means of topographic survey and bathymetric surveys, Samil Beach, Spain. *Shore Beach* **2002**, *70*, 3–10.

Journal of
Marine Science and Engineering

MDPI

Article

How Effective Were the Beach Nourishments at Cancun?

Raúl Martell [1], Edgar Mendoza [2,*], Ismael Mariño-Tapia [3], Itxaso Odériz [2] and Rodolfo Silva [2]

1 Comisión Nacional para el Conocimiento y uso de la Biodiversidad, CONABIO, Mexico City 14010, Mexico; rmartell@conabio.gob.mx
2 Instituto de Ingeniería, Universidad Nacional Autónoma de México, Mexico City 04510, Mexico; itxaso.oderiz@gmail.com (I.O.); rsilvac@iingen.unam.mx (R.S.)
3 Escuela Nacional de Estudios Superiores Unidad Mérida, Universidad Nacional Autónoma de México, Mérida, Yucatán 97357, Mexico; imarino@enesmerida.unam.mx
* Correspondence: emendozab@iingen.unam.mx; Tel.: +52-55-5623-3600

Received: 4 May 2020; Accepted: 26 May 2020; Published: 28 May 2020

check for updates

Abstract: Beach nourishment is generally seen as the preferred means of rectifying coastal erosion, due to its low environmental impact and natural evolution. The largest beach nourishment project ever carried out in Mexico took place on Cancun beach in 2006, as a response to the most intense hurricane season ever registered in Mexico, in 2005. After Hurricane Dean, in 2009, a second nourishment was conducted, which evidenced flaws in the design and execution of the first project. Previous investigations report that the need for beach re-fills directly correlates with wave energy. However, following a thorough revision of the extreme climatic events that occurred between 1978 and 2018, it has been found that the amount of erosion also depends on the frequency and duration of high energy events. The findings also show that the apparent success of the second nourishment is mainly associated with a decline in the number of extreme wave power events impacting the beach. In the conclusion to this paper, we share the knowledge gained, but not yet applied, in Mexico or elsewhere, regarding beach use, urbanization, and protection in beach planning.

Keywords: beach nourishment; Cancun beach; coastal erosion; hurricane damage; beach nourishment assessment; beach profile imbalance

1. Introduction

It has long been recognized that sandy beaches around the world are being eroded, e.g., [1,2] estimated that 70% of beaches were undergoing erosion and, according to [3], the problem was much more critical in the USA, where the figure was 90%. In the Caribbean Basin, [1,4,5] have estimated severe erosion rates for various beaches. In areas where there is potential for tourist or urban development, and erosion problems are detected, scientifically based engineering solutions are expected to control or mitigate these phenomena, e.g., [6].

Over time, the methodologies used for these solutions have evolved, e.g., [7]. In the past, the most frequently applied techniques were based on hardening the coast, by means of dikes, breakwaters, groins, etc. Despite initial criticism, the use of softer solutions (i.e., artificial beach nourishment) has gained popularity, becoming the preferred alternative for mitigating erosion [8–10]. The documented benefits [11–13] of artificial nourishment include:

- The extension of the berm and beach profile, combined with protective dunes, dissipates wave energy, thus reducing expected damage from storms.

- The berm is aesthetically valuable for tourism.
- The extension of the beach increases its lifespan by delaying long-term chronic erosion.
- The profile gradually re-shapes to fit the hydrodynamic conditions of the area, promoting stable conditions.
- If the correct source of sediment is chosen (sorting, shape, quality and density), the nourishment will not induce drastic changes to the water current circulation patterns: color, transparency, temperature, pH, Dissolved Oxygen, Biological Oxygen Demand, Biochemical Oxygen Demand. Nor will it impact organisms, such as nekton, plankton and benthos.
- Structural erosion (scouring) is avoided.
- Implementation and maintenance costs are lower than those corresponding to rigid engineering works.

These benefits mean that when a beach has a sand deficiency, is important in terms of economic impact and job creation, and the environmental impacts are not very serious, beach nourishment can be seen as a feasible alternative to remedy problems of erosion.

In 1923, over a million cubic meters of sand were placed on a beach near Coney Island, New York, making this the first cited case of artificial beach nourishment in modern times [13,14]. From then until the 1950s some 72 beaches in the USA were artificially nourished. However, not all of these projects were based on scientific grounds, and most were carried out empirically. Many of the projects had a short life, arguably due to the inappropriate selection of the sediment used. In the United States and Europe, from the 1950s to the 1970s, many authors, including [15–19], offered technical criteria for the design of beach nourishment projects. These works formed the scientific basis for various compilations, such as [8,20–23], from which most artificial sand nourishment projects worldwide have taken their evidence.

In Europe, the Netherlands leads the field in coastal protection, with 50% of its territory being below sea level. In 1984 the Netherlands introduced regulations for coastal protection based on artificial nourishment of their beaches and sand dunes [24]. The research carried out there and the experience accumulated confirmed the functional efficiency of this technique, as well as its adaptability and lower costs, compared to alternative methods [25]. Researchers in the Netherlands have contributed considerably to the knowledge of the criteria for designing sand nourishment projects, including the dune, berm, and submerged beach. Between 1952 and 1989, 60 million cubic meters of sand were placed on the coast of the Netherlands in over 50 projects. From the experience gained, in 1990 the Dutch government stated that beach nourishment was the first option in adapting to climate change, and an innovative large-scale beach nourishment project was begun, using the so-called Sand Engine strategy [26].

Important artificial nourishment projects have also been executed in Spain, Germany, the United Kingdom, France, Portugal, Georgia, Japan, and Australia. In the Caribbean area, some projects in Mexico, Cuba, and the Dominican Republic stand out, because of the large amount of biogenic sand required in the nourishments.

As with any engineering work, beach nourishment projects have a limited lifetime. This lifespan depends upon the sediment transport induced by the power and persistence of extreme sea states and the resilience of the system [27]. According to [28], an artificially nourished beach will have to be re-nourished, depending on the characteristics of its design and evolution. Artificial sand nourishment is simply a mitigation procedure, which needs periodic maintenance. If the environmental conditions are favorable, infilling intervals can be extended, with corresponding financial savings. On the other hand, it is accepted that if the intensity and frequency of the storms in the area increase, the erosion rates will also rise, as will associated costs of sand refills, perhaps making the whole procedure unsustainable. However, together with the intensity, the duration of heavy sea states is a variable which determines the lifespan of a beach nourishment. In any case, alternative measures should always be considered, including the use of rigid structures. If that also fails, a managed retreat of the infrastructure would be the only alternative.

Assessing the effectiveness of sand nourishment projects is important from both economic and technical perspectives. This is done through a monitoring program, by periodically measuring the morphological changes of the beach. The aim of this work is to assess the effectiveness of two beach nourishments carried out at Cancun, Mexico, in 2006 and 2009, using a total of 8 million cubic meters of sand. This assessment uses satellite images, photographs, and model-derived wave power records to examine the morphological evolution of the coastline over the last 40 years.

2. Cancun Before the First Beach Nourishment

Cancun is located on the Caribbean coast of the Yucatan Peninsula, in Mexico (Figure 1). The beach is a prime example of an environment which has been progressively degraded by human activities linked to tourism. The urbanization of Cancun started in the late 1960s on a pristine barrier island between the Caribbean Sea and a system of coastal lagoons (Nichupté). The lagoons are supplied with fresh water, via continental ground water discharge, and with sea water, which arrives with the tides (micro-tidal regime), through two inlets. The lagoons used to be surrounded by a very rich mangrove forest, which is now greatly reduced. The beach of Cancun lacks the protection of coral reefs and the sediment is composed mostly of biogenic sediments.

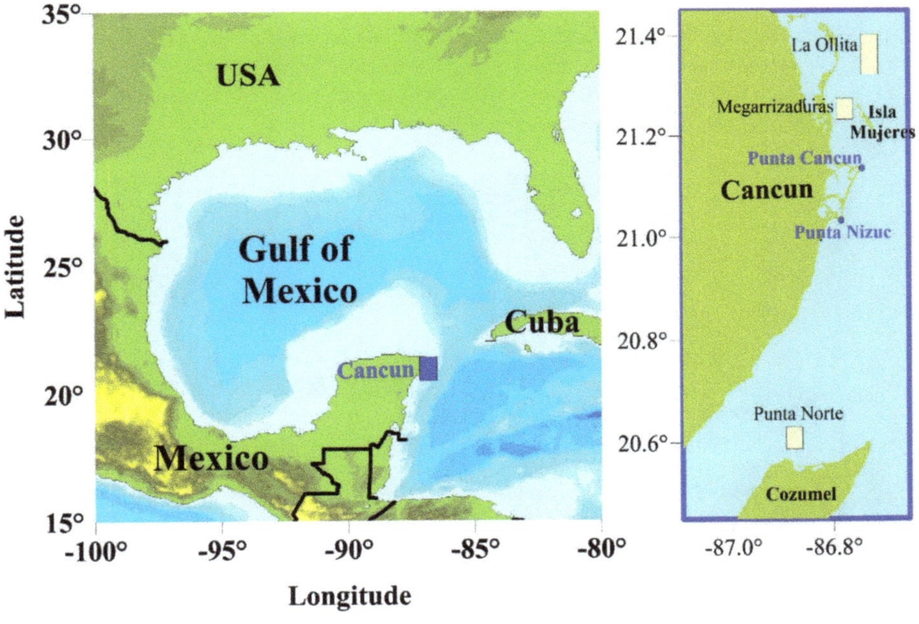

Figure 1. Location of the barrier island beach of Cancun, between Punta Cancun and Punta Nizuc.

The main anthropic alteration to the Cancun coast has been the construction of very dense infrastructure (hotels, roads, gardens, golf courses, etc.) on the dunes. The inlets were also rigidized, causing modifications to the natural breaching of the sandbar during storms, and thus limiting the interaction between the sea and lagoon. These changes, together with a decline in the amount of natural sediment availability, brought about intense erosion. Beach loss was estimated at an annual rate of 1.8 m for 1967–2005 [29]. Figure 2 shows Cancun in 1947, and the slow, but continuous urbanization of the beach between 1978 and 1988.

Figure 2. Historical aerial images of Cancun. Punta Cancún in 1947 (left panel) and Punta Nizuc in 1978 and 1988 (center and right panels, respectively).

As documented by [30] and [31], until 1988 the urban planning of Cancun did not take into consideration the natural dynamics of the coast. In September 1988, Cancun received the full force of hurricane Gilbert. The waves and storm surge moved massive quantities of sand and, for the first time, the hoteliers there experienced the effects of substantial erosion on the beach. Several studies were carried out, but no action was taken to recover the beach. The individual hotel owners implemented uncoordinated, inadequate coastal protection schemes (e.g., vertical walls, detached breakwaters, and groins), accelerating the problem and transferring the effects to neighboring beach segments (see Figure 3). Even so, the beach partially recovered until hurricanes Ivan (2004) and Emily (2005) hit the coast [32]. Then, in October 2005, Hurricane Wilma hit Cancun, removing 8 million m^3 of sand [33]; almost all of the beach was left without sediment (Figure 4).

Figure 3. Examples of inadequate solutions implemented 1988–2004. Vertical walls (panels **A** and **B**); detached structures (panel **C**) and groins (panel **D**).

Figure 4. Examples of damage caused by Hurricane Wilma to the beach of Cancun in 2005. It can be seen that in front of many hotels the beach totally disappeared; only rocks were left behind (panels (**a**), (**c**) and (**f**)), while in other places, waves were hitting the hotels walls and facilities directly (panels (**b**), (**d**) and (**e**)).

3. The First Beach Nourishment, 2006 to 2009

In 2006, the problems of erosion had come to a head; the situation was critical economically as the all-important tourist industry in Cancun was in jeopardy. An urgent solution that could restore the beach immediately and activate the economy was necessary and a beach nourishment was undertaken to achieve this goal. From January to April 2006, 2.7 million cubic meters of sand were placed on the beach, with a cost of USD 19 million. The sediment was borrowed from two nearby sand banks of La Ollita and Megarrizaduras (Figure 1). La Ollita sand bank is 12 km north-east of Cancun, off the northern end of Isla Mujeres. Around 1.7 million cubic meters of sand were extracted from here, from an average depth of 25 m [33]. The Megarrizaduras sand bank is 15 km north of Punta Cancun, in shallow waters (7–10 m depth) between Cancun and Isla Mujeres. Around one million cubic meters of sand were extracted from this bank.

Immediately after the sand nourishment, the beach had an average width of 60 m (Figure 5). However, the beach rapidly began to narrow, losing an average 8 m in width, from May to September 2006, and even more, 29 m, from September 2006 to September 2007 [34]. In the same period, the sand on the beach formed a dramatic, pronounced scarp, making it very difficult for tourists to enjoy the beach (see upper panels of Figure 6). These scarps remained, due to the lack of sediment in the submerged part of the beach profile. The beach was dramatically damaged by hurricane Dean in August 2007.

Figure 5. Aerial views (from north to south) of various segments of the central part of Cancun beach in May 2006; the immediate result obtained from the nourishment is evident.

Figure 6. Examples of damage caused by Hurricane Dean to the beach of Cancun. Upper panels: August 15, 2007 (a few days before Hurricane Dean hit land), and lower panels: August 25, 2007 (a few days after Hurricane Dean had landed). Each photograph corresponds to approximately the same place in the upper and lower panel.

By 2007 the level of anthropization of the barrier island was over 95%, with commercial centers, hotels, houses, gardens, and a wide boulevard covering almost all the areas that had originally been dunes, coastal vegetation, and mangrove (Figure 7 [35]).

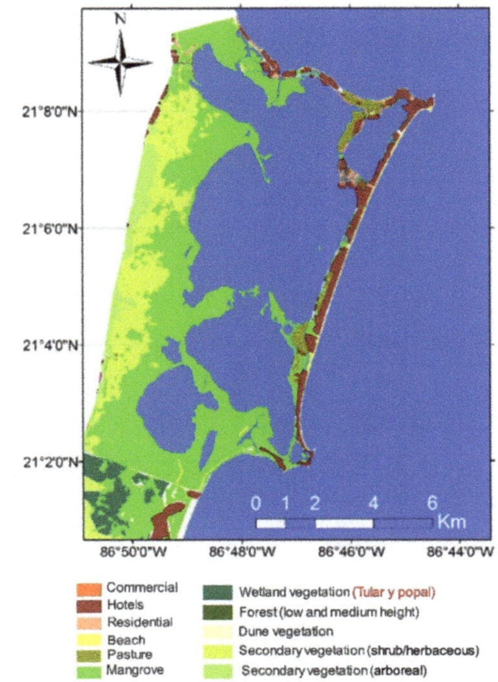

Figure 7. Land use, Cancun 2007 [35]. Intense urbanization all along the barrier rigidized the coast and facilitated the loss of beach sediment.

On 21 August 2007, Hurricane Dean hit Cancun, generating very intense waves that removed significant volumes of sand; the dry beach was reduced in width and the scarps disappeared (Figure 4). After Hurricane Dean, the erosion on the beach continued, with a recorded loss of 10 m of dry beach, from September 2007 to August 2008, and a further 1 m between August 2008 and January 2009 [36]. Although the rate of the erosion diminished, by January 2009 the waves were lapping only 14.5 m from the sea walls of the hotels [37]. The erosion continued, bringing a shoreline recession in the central and northern sections of the beach to a position similar to the post-Wilma conditions of 2005.

From exhaustive monitoring of the beach evolution in 2006–2009, [38] concluded that during extreme wave conditions, longitudinal currents are highly dynamic and determine sand availability at the northern and southern ends of Cancun beach. In the central part of the system, offshore sediment transport is dominant and the sediment balance is negative. This sediment imbalance prevents the natural self-regulation of the coastal system; causes very low sediment input into the system; and leads to large amounts of sediment being transported out of the system by hurricane-induced high wave energy. These conclusions have been very important in understanding the beach behavior at Cancun.

4. The Beach Nourishment of 2010, until 2020

Given that the beach had not reached a dynamic equilibrium, it continued eroding and consequently was not attractive for tourists (Figure 8). A second beach nourishment was therefore performed, from December 2009 to January 2010. The methodology was similar to that of 2006. However, this time the amount of sand used to replenish the beach was almost twice that of 2006 (5.2 million cubic meters). The width of the beach at the end of the process was 80 m and the thickness of the sand layer in the beach profile was substantially higher than that of 2006 (Figure 9). Additionally, a 305 m long breakwater crowned 2.5 m above the mean sea level was built on Punta Cancun, to prevent the transport of the sediment to the north.

Figure 8. View of the beach at Cancun in 2009, before the second nourishment. Waves attacking the hotel facings (left) and rocks left behind after erosion (right).

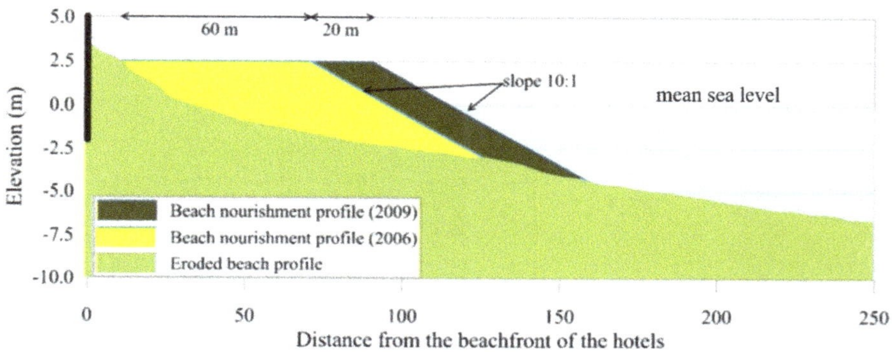

Figure 9. Projected profiles for the 2006 and 2009 beach nourishments.

For the second nourishment project, 2.5 million cubic meters of sand were extracted from La Ollita II sand bank, next to the sand bank used in 2006. The sand was extracted from an average depth of 25 m. The remaining 2.8 million cubic meters of sediment were taken from the sandbank at Punta Norte, in the shallow waters north of the island of Cozumel, 48 km away, at a depth of 12–29 m [33]. While all the borrowed sand was of marine biogenic origin, the sand used had different mechanical characteristics from the sand native to Cancun beach (see Figure 10). The differences in the sand characteristics may explain the unpredictable behavior of sand transport on the beach and therefore the beach evolution after the second nourishment.

Figure 10. Left panel shows the native sand, 2004. Central and right panels, sand from the emerged and submerged beach fill, respectively.

The estimated lifespan of the second nourishment was 3 years [33], a period calculated on the basis of the trends in severe storm impacts previously observed. Nevertheless, after the beach nourishment process, the system never attained the expected profile (see Figure 11). The wave action caused a steep beach scarp to form; an unpleasant 2 m vertical step (Figure 12). A very thorough analysis of this process on the beach in Cancun was made by [38].

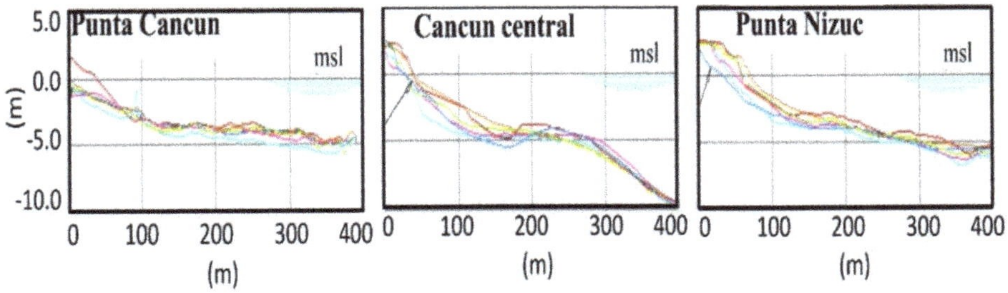

Figure 11. Beach profiles at northern (P. Cancun); central, and southern (P. Nizuc) Cancun, showing the beach growth due to the first beach fill (early 2006), and a return to erosion conditions by 2008, when the need for a 2nd nourishment was evident: Nov-2005 (– –), May-2006 (– –), Aug-2006 (– –), Apr-2007 (– –), Sep-2007 (– –) and Aug-2008 (– –).

Figure 12. The second beach nourishment was carried out in June 2010. By October 2010 the beach had developed a scarp of almost 2 m height as seen in both photographs.

In 2013 all the scarps had disappeared (Figure 13) and, 10 years after the second nourishment, the beach at Cancun has managed to conserve a nearly stable width of approximately 30 m (Figures 14 and 15). Therefore, no further beach nourishments have been required.

Figure 13. Aspect of the beach of Cancun in 2013. All the scarps have disappeared and the beach seems to be in a sound state.

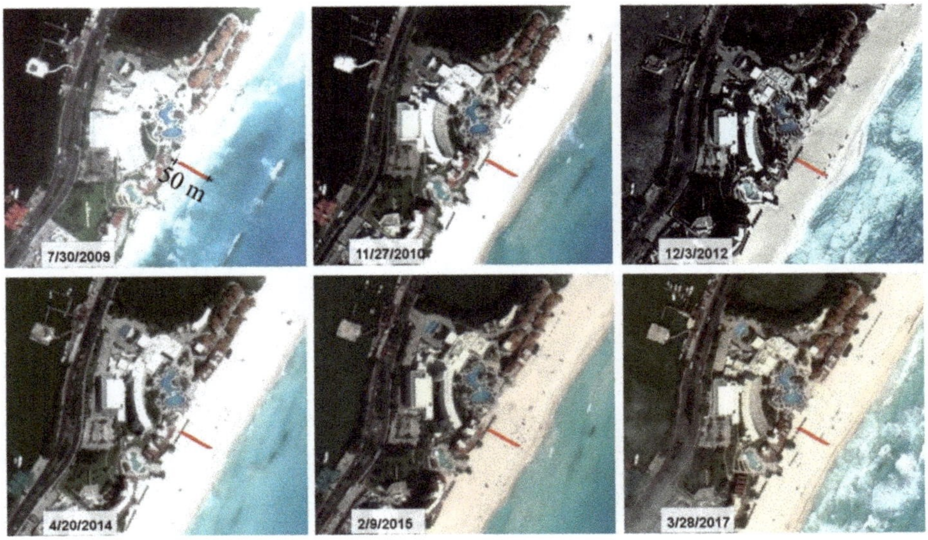

Figure 14. Satellite images, 2009 to 2017, of the beach in front of the Grand Park Royal Cancun, located in the central-north part of the beach (source: Google Earth Pro, version: 7.3.2.5776). The red line, 50 m long, shows the dynamic stability being reached.

Figure 15. Recent views of the beach at Cancun, January 16, 2020 [39].

5. Characterization of Extreme Events 1979–2018

To evaluate the response of the beach nourishment to the wave climate, three periods, namely Stage 0 from 1979 to 2005; Stage 1 from 2006 to 2009; and Stage 2 from 2009 to 2018, were analyzed separately. For this characterization, wave records for 21.0 °N and 86.5 °W were used. Hourly information of significant wave height (H_s), peak wave period (T_p), mean wave period (T_m), and wave direction (θ) were taken from the reanalysis dataset ERA 5 [40] of the European Centre for Medium-Range Weather Forecasts (ECMWF).

An extreme event was defined as when the significant wave height exceeds the general 90th percentile significant wave height (H_{s90}) [41]. This value was 2.10 m for Stages 0 and 1, and 2.03 for Stage 2. If the difference in time between the events is more than 12 h, these are considered independent events [42]. The deep-water parameters for each extreme event found of the wave time series were plotted against the number of events, that is: event *duration*, storm cluster time intervals (*IN*), significant wave height (H_s), maximum significant wave height (H_{smax}), peak period (T_p), maximum peak period (T_{pmax}) and mean wave power by event (P_{wE}). The storm cluster *IN* is the time elapsed (days) between storms [43].

The results shown in Figure 16 give an overview of the intensity, persistence, and clustering of the extreme events which occurred in each of the three stages. All of these factors are important in the morphodynamic response of the beach. In Figure 16 the variables are presented in histograms, to facilitate comparisons between the stages.

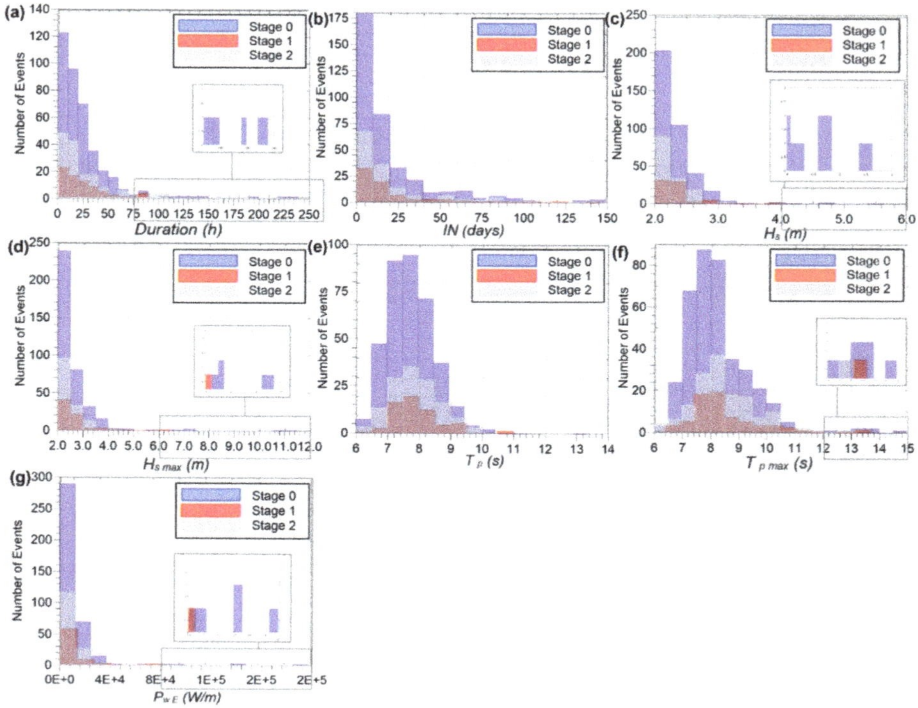

Figure 16. Histograms showing the extreme wave parameters for Stage 0 (blue), Stage 1 (red), Stage 2 (grey); (**a**) Duration; (**b**) *IN* (storm cluster time intervals); (**c**) H_s (significant wave height); (**d**) $H_{s\ max}$ (maximum significant wave height); (**e**) T_p (peak period); (**f**) $T_{p\ \max}$ (maximum peak period); (**g**) P_{wE} (mean wave power by event).

Figure 16 shows quite similar distributions of the variables for all three stages. As the period is much longer, Stage 0 shows a greater number of events in all the histograms and also more events in the tail of the distributions; this is a natural phenomenon. The duration of the individual events shows no significant change between the stages. Of particular interest is the low values of *IN* found in Stage 2, which show that those storm seasons were more active and, when this time is less than the natural recovery period of the beach, the damage to the beach tends to accumulate. Panels (**c**), (**d**), (**e**) and (**f**) show that the most intense storms occurred in Stages 0 and 1, whilst in Stage 2 the reduction in the number and intensity of storms is clear.

These findings, analyzed for each stage, are in agreement with the levels of wave power (intensity, persistence) generated by the most intense events. From the information in Figure 16, the levels of erosion reported by [44] can be more clearly understood: in Stage 0 from 1983 to 1990, the eroded area was 218,000 m^2; during the 1990s, the erosion was lower; 19,600 m^2, while the eroded surface from April to October 2005 was 265,000 m^2. The hydrodynamic characteristics of extreme events and other factors, such as translation speed and trajectory, sea level, wind effects, and sediment properties, contribute to beach erosion. However, it has been reported elsewhere that episodic extreme storms and a series of lesser storms occurring close together could result in greater coastal impact than one single huge storm [45–47]. This means that longer periods of persistently high wave energy can lead to great changes in the beach profile [48,49]. Therefore, in this paper, a detailed analysis focusing only on wave power was performed.

For the study of the wave power, a time series was calculated using the formula for irregular waves (Equation (1)), see Figure 17:

$$P_W = \frac{\rho g^2}{64\pi} H_s^2 T_e \ (W/m). \quad (1)$$

where H_s (m) is the significant wave height and T_e is the energy period of the spectra, which can be expressed as a linear function of the mean wave period T_{01}(s),

$$T_e = \frac{m_{-1}}{m_0} = \alpha T_{01}; \quad (2)$$

where $\alpha = 1.08$ [43].

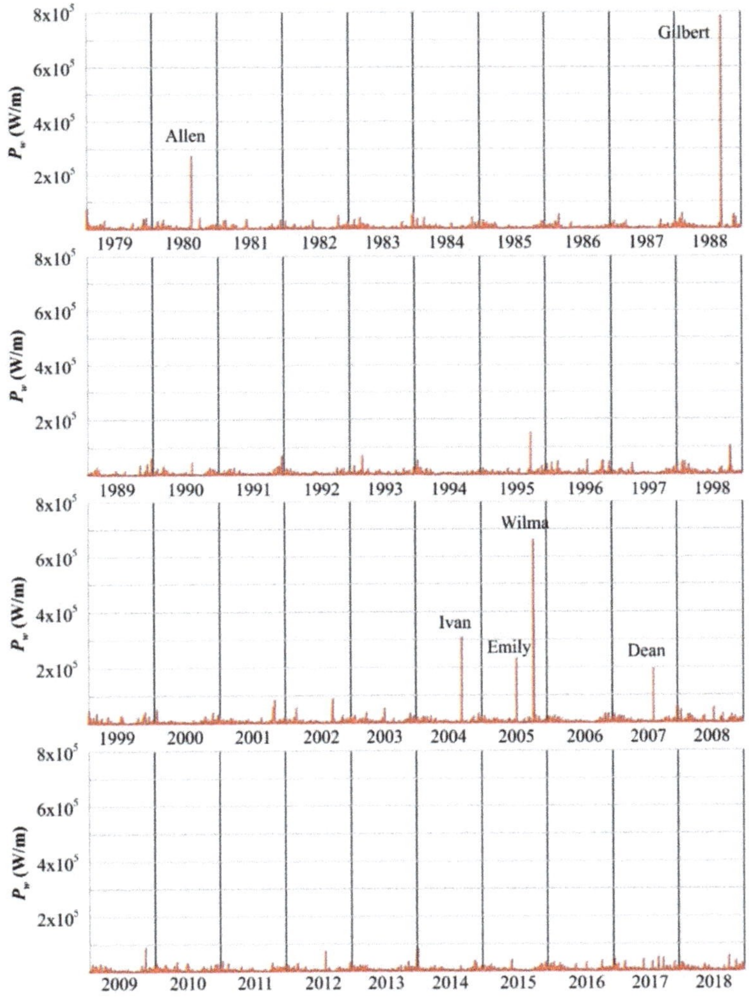

Figure 17. Hourly wave power at 21.0 °N and 86.5 °W.

From 1979 to 2018, six events reached or exceeded a maximum wave power of 2×10^5 W/m (see Figure 17); this threshold was selected, being the greatest erosion reported (general knowledge).

These hurricanes were named Allen (1980), Gilbert (1988), Ivan (2004), Emily (2005), Wilma (2005), and Dean (2007). Their characteristics are presented in Table 1. According to reports, these six hurricanes generated the waves which induced the most severe beach erosion. However, the orders of magnitude of the sand eroded were not directly correlated with the maximum wave power reached in these events. Looking at the dates of the greatest events (Figure 17), and considering their durations (Table 1), together with the largest beach losses reported, erosion seems to be closely related to the intensity and duration of extreme sea states.

Table 1. Wave characteristics of the most extreme storm conditions induced by the hurricanes from 1979 to 2018.

Event	Allen	Gilbert	Ivan	Emily	Wilma	Dean
Date over 10^4 W/m	07/08/1980	13/02/1988	12/09/2004	17/07/2005	19/10/2005	20/08/2007
Date P_{wmax}	07/08/1980	14/09/1988	13/09/2004	18/07/2005	21/10/2005	21/08/2007
Duration (h)	48	93	90	34	161	44
$\overline{H_s}$ (m)	3.80	4.30	3.83	3.52	5.19	3.53
H_{smax} (m)	7.43	11.28	7.13	6.73	10.67	6.06
$\overline{T_p}$ (s)	8.46	9.30	12.61	10.40	11.51	9.98
$\overline{\theta}$	37.00	100.56	57.59	108.33	133.21	109.12
P_{wmax} (10^6 W/m)	0.28	0.78	0.31	0.23	0.66	0.20
P_{wtotal} (10^6 W/m)	3.85	12.30	9.51	2.48	29.42	3.23

The arrival direction of the wave power to the shore can limit its erosiveness. For this reason, the dominant direction of the incoming waves from 1979 to 2018 was taken from the same ERA-5 database. The results, separated into Stages 0, 1 and 2, can be seen in Figure 18.

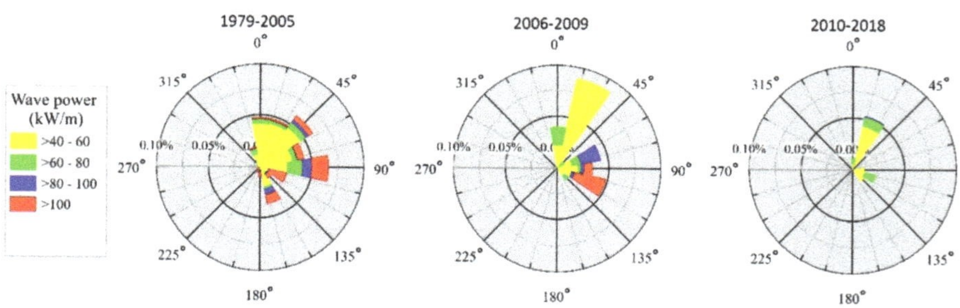

Figure 18. Distribution of the wave power according to incident direction. Stage 0 (1979–2005); Stage 1 (2006–2009); Stage 2 (2010–2018). North direction is 0°.

From Figure 18 it can be inferred that, during Stage 0, the extreme events induced sediment transport to leave the littoral cell in all directions. In Stage 1, most transport was from the south to the north and during Stage 2 no events are found that could produce relevant sediment transport to cause beach loss. The Cancun beach, like any other, has seasonal and interannual variations in erosion-accretion. However, when the wave power exceeds a threshold of around 100 kW/m, the sand is exported out of the coastal cell and the system is not able to recover naturally

6. Discussion and Conclusions

Between 1988 and 2012, six major hurricanes impacted Cancun and information was demanded. As a result, a large number of coastal and oceanographic studies were undertaken, focusing on Cancun, and paying attention to developing human and technical capabilities. Until 2010, the time of the second nourishment, the general perception was that the beach at Cancun had a clear erosion tendency. In this section we will show that this tendency was directly linked to the incidence of extreme hydrodynamic conditions and the scarcity of natural sediment sources. These findings could change erroneous perceptions, and present many challenges for future nourishment projects, includes the need for improving long term predictions of wave climate under global warming scenarios.

Previous research [37,38] explained the dynamics of Cancun beach in terms of wave energy fluxes with reasonable accuracy. Therefore, in this paper we opt for a long-term analysis, only evaluating the overall wave power that reaches the beach. With this information, we identified the events which produced the most erosion at Cancun beach. The effects of these climatological events were identified by a cluster analysis which suggested three stages, before the first nourishment (1979–2005), after this (2005–2009), and after the second nourishment (2010–2018). As storms were more frequent in Stages 0 and 1, the recovery capacity of the beach was debilitated.

The analysis of wave power in the region indicates that for the periods before and after the 2010 sand nourishment, the severity of the hydrodynamic conditions was very different. In the period 1988–2009, the beach was repeatedly affected by intense storms, which induced very high erosion rates. In the period after the second beach nourishment (2010–2018), storm impact was very low. The analysis shows that for the last 8 years, no major hurricanes have impacted the region, and the beach of Cancun has been stable. In this sense, beach stability means that Cancun has maintained a beach width suitable for recreation and for the protection of infrastructure from inundation and damage (i.e., >40 m wide) for most of its length, most of the time (see Figure 14). In the absence of hurricanes, all work related to beach monitoring has ceased, despite the importance of Cancun in terms of the tourist industry. Decision makers and investors believe that the beach is stable and that the future of this resort was guaranteed with the second beach nourishment. However, Cancun has taught us that sudden erosion processes occur when we are least prepared.

The considerable amount of sand deposited on the beach (8 million cubic meters), along with the low storm impact in the last 13 years, has led the beach away from an erosion tendency. However, it is still too early to assert whether this balance will continue over time, given that the main causes of the erosion are still present.

In the last 50 years of touristic development at Cancun, the exploitation of the dunes, and the use of the beach for recreation, there have been various lessons that should have been learnt:

1. The passage of Hurricane Gilbert in 1988 showed that the resulting erosion put at great risk the touristic infrastructure.
2. Between 1988 and 2004, it was seen that without the presence of very severe waves the beach can slowly and partially recover.
3. During this period, we also learned that the individual "coastal protection" efforts (by hoteliers) in fact increased structural erosion and that such uncoordinated efforts should be avoided. The need for comprehensive coastal management plans and solutions for the entire beach was made evident.
4. The number of nearby sand banks available for borrowing sand is limited; these areas cannot be exploited indefinitely. As this sand is of biogenic origin, the production of these sediments is a very long-term process.
5. Beach scarps are unsuitable for tourism. The design and techniques employed for beach nourishment has to be different, since for long periods the beach profile had a 2 m scarp.
6. From the morphodynamic behavior of neighboring beaches in the same period, it is evident that the impact of urbanization on dune ecosystems reduces their resilience.
7. The experience gained from the first beach nourishment and subsequent coastal monitoring programs, implemented by government and research bodies, enhanced the design of the second beach nourishment.
8. Much to the contrary of what was previously thought, it has been seen that, in the last 10 years, the beach has not gradually eroded. The erosion events are episodic and the beach may recover again. Nevertheless, it is expected that an extreme event could drive Cancun rapidly to an

unstable stage, since much of its natural resilience has been lost due to the urbanization of the coastal dunes.

Cancun is a vital asset to the economy of Mexico. In order to have a tourist destination with a healthy sustainable beach, a series of actions are necessary, including the re-establishment of coastal monitoring programs to have more accurate information to correctly diagnose the beach behavior. It is also important to continue comprehensive coastal planning work that allows the coexistence of natural processes and sea-sand-sun tourism.

Author Contributions: Conceptualization, R.M., R.S. and E.M.; methodology, I.M.-T., R.S. and I.O.; software, I.O.; validation, R.S., I.M.-T. and E.M.; formal analysis, E.M. and R.S.; investigation, R.S. and I.O.; resources, R.S. and I.M.-T.; data curation, I.O. and R.S.; writing—original draft preparation, R.M., E.M. and I.M.-T.; writing—review and editing, R.S., E.M. and I.M.-T.; visualization, R.S., E.M. and I.O.; supervision, R.S. and I.M.-T.; project administration, R.S. and I.M.-T.; funding acquisition, R.S. and E.M. All authors have read and agreed to the published version of the manuscript.

Funding: This research received no external funding.

Acknowledgments: The authors wish to thank CONACYT Mexico and CYTED project "Protección de Frentes Urbanos Costeros Frente al Calentamiento Global" under contract P916PTE0234 for partially funding this research.

Conflicts of Interest: The authors declare no conflict of interest.

References

1. Silva, R.; Martínez, M.L.; Hesp, P.A.; Catalan, P.; Osorio, A.F.; Martell, R.; Fossati, M.; Miot da Silva, G.; Mariño-Tapia, I.; Pereira, P.; et al. Present and future challenges of coastal erosion in Latin America. *J. Coast. Res.* **2014**, *71*, 1–16. [CrossRef]
2. Bird, E.C.F. *Coastline Changes-A Global Review*; John Wiley-Interscience: Chichester, UK, 1985; p. 219.
3. Leatherman, S.P. Beach Response Strategies to Accelerated Sea Level Rise. In Proceedings of the 2nd North American Conference on Preparing for Climate Change, Washington, DC, USA, 6–8 December 1988; pp. 353–358.
4. Cambers, G. Coastlines of the Caribbean. In Proceedings of the 7th Symposium on Coastal and Ocean Management, Long Beach, CA, USA, 8–12 July 1991; ASCE: New York, NY, USA, 1991; p. 187.
5. Escudero, M.; Felix, A.; Silva, R.; Marino-Tapia, I.; Mendoza, E. Beach erosion and loss of protection environmental services in Cancun, Mexico. *Ocean Coast. Manag.* **2018**, *156*, 183–197. [CrossRef]
6. Silva, R.; Lithgow, D.; Esteves, L.; Martínez, M.L.; Moreno-Casasola, P.; Martell, R.; Pereira, P.; Mendoza, E.; Campos-Cascaredo, A.; Winckler-Grez, P.; et al. Coastal Risk Mitigation by Green Infrastructure in Latin America. *Proc. Inst. Civ. Eng.-Mar. Eng.* **2017**, *170*, 39–54. [CrossRef]
7. Silva, R.; Chávez, V.; Bouma, T.J.; van Tussenbroek, B.I.; Arkema, K.K.; Martínez, M.L.; Oumeraci, H.; Heymans, J.J.; Osorio, A.F.; Mendoza, E. The incorporation of biophysical and social components in coastal management. *Estuar. Coast* **2019**, *42*, 1695–1708. [CrossRef]
8. Dean, R.G.; Dalrymple, R.A. *Coastal Processes with Engineering Applications*; Cambridge University Press: Cambridge UK, 1988; p. 489.
9. Stive, M.J.F.; De Schipper, M.A.; Luijendijk, A.P.; Ranasinghe, R.W.M.R.J.B.; Van Thiel De Vries, J.S.M.; Aarninkhof, S.; Marx, S. The Sand Engine: A solution for vulnerable deltas in the 21st century? In Proceedings of the Coastal Dynamics 2013: 7th International Conference on Coastal Dynamics, Arcachon, France, 24–28 June 2013.
10. Escudero, M.; Mendoza, E.; Silva, R. Micro sand engine beach stabilization strategy at Puerto Morelos, México. *J. Mar. Sci. Eng.* **2020**, *8*, 247. [CrossRef]
11. Dean, R.G. Additional Sediment Input to the Nearshore Region. *Shore Beach* **1987**, *55*, 76–81.
12. CUR. *Manual on Artificial Beach Nourishment*; Codes and Specifications, Repot 130; Centre for Civil Engineering Research: Gouda, The Netherlands, 1987; p. 195.
13. Davison, A.T.; Nicholls, R.J.; Laetherman, S.P. Beach Nourishment as a Coastal Management Tool: An Annotated Bibliography on Developments Associated with the Artificial Nourishment of Beaches. *J. Coast. Res.* **1992**, *8*, 984–1022.

14. Walker, H.J.; Finkl, C.W. Beach nourishment: Case studies. In *Engineered Coasts*; Kluwer Academic Publishers: Dordrecht, The Netherlands, 2002; pp. 23–60.
15. Bruun, P. *Coast Erosion and the Development of Beach Profiles*; Tech. Memo No. 44; US Beach Erosion Board: Washington, DC, USA, 1954; p. 79.
16. Bagnold, R.A. *Mechanics of Marine Sedimentation*; Sea, M.N., Ed.; Interscience: New York, NY, USA, 1963; pp. 507–528.
17. Komar, P.D. *Beach Processes and Sedimentation*; Prentice Hall: Upper Saddle River, NJ, USA, 1973; p. 544.
18. Dean, R.G. Heuristic Models of Sand Transport in the Surf Zone. In Proceedings of the Conference on Engineering Dynamics in the Surf Zone, Sydney, Australia, 1973; pp. 208–214.
19. James, W.R. Beach Fill Stability and Borrow Material Texture. In Proceedings of the 14th ICCE, Copenhagen, Denmark, 24–28 June 1974; ASCE: Reston, VA, USA, 1974; pp. 1334–1349.
20. Van Rijn, L.C. *Principles of Coastal Morphology*; Aqua Publications: Amsterdam, The Netherlands, 1988.
21. Herbirch, J.B. *Handbook of Coastal Engineering*; McGraw-Hill: New York, NY, USA, 2000.
22. Woodroffe, C.D. *Coast: Form, Process and Evolution*; Cambridge University Press: Cambridge, UK, 2002.
23. Dean, R.G. *Beach Nourishment: Theory and Practice*; World Scientific Publishing Co. Pte. Ltd.: Singapore, 2002; p. 420.
24. TAW. *Guidelines for the Evaluation of Safety of Dunes as Coastal Defense (English Translation of the Official Dutch Guidelines)*; CUR Publishing Foundation: Gouda, The Netherlands, 1986; p. 26.
25. Van de Graaff, J. Probabilistic design of dunes: An example from The Netherlands. *Coast. Eng.* **1986**, *9*, 479–500. [CrossRef]
26. Van Slobbe, E.; de Vriend, H.J.; Aarninkhof, S.; Lulofs, K.; de Vries, M.; Dircke, P. Building with Nature: In search of resilient storm surge protection strategies. *Nat. Hazards* **2013**, *66*, 1461–1480. [CrossRef]
27. Martínez, M.L.; Taramelli, A.; Silva, R. Resistance and resilience: Facing the multidimensional challenges in coastal areas. *J. Coast. Res.* **2017**, *77*, 1–6. [CrossRef]
28. National Research Council. *Beach Nourishment and Protection*; National Academy Press: Washington, DC, USA, 1995.
29. Martell, R.; Mariño, I.; Mendoza, E.; Silva, R. Variaciones morfológicas a largo plazo del perfil de playa en Cancún, México. In Proceedings of the XXI Congreso Nacional De Hidráulica, Jalisco, México, 25 November 2010; p. 8.
30. Mendoza, E.; Silva, R.; Enriquez-Ortiz, C.; Mariño-Tapia, I.; Felix, A. *Analysis of the Hazards and Vulnerability of the Cancun Beach System, in Extreme Events: Observations, Modeling, and Economics*; Chavez, M., Ghil, M., Urrutia-Fucugauchi, J., Eds.; John Wiley & Sons, Inc.: Hoboken, NJ, USA, 2015; pp. 125–136.
31. Villatoro, M.M.; Escudero, M.C.; Mendoza, E.; Silva, R. Chapter 7.7. Balancing flood and erosion risk with landscape sustainability: Cancun, Mexico. In *Coastal Risk Management in a Changing Climate*; Zanuttigh, B., Nicholls, R., Vanderlinden, J.-P., Thompson, R., Burcharth, H., Eds.; Elsevier: Amsterdam, The Netherlands, 2014; pp. 506–534.
32. Martell, R.; Mariño, I.; Mendoza, E.G.; Silva, R. Variaciones morfológicas, inducidas por condiciones hidrodinámicas extremas, en la playa de Cancún, México. XXV Congreso Latinoamericano de Hidráulica, San José, Costa Rica, 12 September 2012; p. 10.
33. CFE. Restauración, recuperación, sostenimiento y mantenimiento de la zona federal marítimo terrestre de Cancún, Playa del Carmen y Cozumel; Manifestación de impacto Ambiental, presentada a SEMARNAT. 2009; 120.
34. Silva, R.; Mariño, I.; Enriquez, C.; Mendoza, E.; Escalante, E.; Ruiz, F. *Monitoring Shoreline Changes at Cancun Beach, Mexico: Effects of Hurricane Wilma*; Word Scientific Publishing, ICCE Coastal Engineering: San Diego, CA, USA, 2006; pp. 3491–3503.
35. Félix, A. Análisis de la dinámica geomorfológica de la zona hotelera de Cancún como contribución al desarrollo de un Plan de Manejo Costero. MSc Thesis, Universidad Autónoma de Campeche, Campeche, México, 2007.
36. Martell, R.; Mariño, I.; Mendoza, E.; Silva, R. Análisis del oleaje swell y sea generado por los huracanes Wilma (2005) y Dean (2007). In Proceedings of the XXII Congreso Nacional de Hidráulica, Guerrero, México, 12 September 2012; AMH: Adelaide, Australia, 2012; p. 8.

37. González-Leija, M.; Mariño-Tapia, I.; Silva, R.; Enriquez, C.; Mendoza, E.; Escalante-Mancera, E.; Ruíz-Rentería, F.; Uc-Sánchez, E. Morphodynamic Evolution and Sediment Transport Processes of Cancun Beach. *J. Coast. Res.* **2013**, *29*, 1146–1157. [CrossRef]
38. Ruiz de Alegria-Arzaburu, A.; Mariño-Tapia, I.; Silva, R.; Pedrozo-Acuña, A. Post-nourishment beach scarp morphodynamics. *J. Coast. Res.* **2013**, *65*, 576–581. [CrossRef]
39. Webcams de México. Available online: wabcamsdemexico.com (accessed on 15 February 2020).
40. Copernicus Climate Change Service (C3S). *ERA5: Fifth Generation of ECMWF Atmospheric Reanalyses of the Global Climate*; Copernicus Climate Change Service Climate Data Store (CDS): Barcelona, Spain, 2018.
41. Bromirski, P.D.; Cayan, D.R.; Helly, J.; Wittmann, P. Wave Power Variability and Trends across the North Pacific. *J. Geophys. Res. Oceans* **2013**, *118*, 6329–6348. [CrossRef]
42. Harley, M. Coastal Storm Definition. In *Coastal Storms: Processes and Impacts*; Ciavola, P., Coco, G., Eds.; John Wiley & Sons: Hoboken, NJ, USA, 2017; pp. 1–19.
43. Dissanayake, P.; Brown, J.; Wisse, P.; Karunarathna, H. Effects of storm clustering on beach/dune evolution. *Mar. Geol.* **2015**, *370*, 63–75. [CrossRef]
44. Silva, R.; Ruiz-Martinez, G.; Mariño-Tapia, I.; Posada-Vanegas, G.; Mendoza, E.; Escalante-Mancera, E. Manmade vulnerability of the Cancun Beach system: The case of hurricane Wilma. *Clean (Weinh)* **2012**, *40*, 911–919.
45. Coco, G.; Senechal, N.; Rejas, A.; Brian, K.R.; Capo, S.; Parisot, J.P.; Brown, J.A.; MacMahan, J.H.M. Beach response to sequence of extreme storms. *Geomorphology* **2014**, *204*, 493–501. [CrossRef]
46. Dolan, R.; Davies, R.E. Coastal storm hazards. *J. Coast. Res.* **1994**, *12*, 103–114.
47. Karunarathna, H.; Pender, D.; Ranasinghe, R.; Short, A.D.; Reeve, D.E. The effects of storm clustering on beach profile variability. *Mar. Geol.* **2014**, *348*, 103–112. [CrossRef]
48. Silva, R.; Martínez, M.; Odériz, I.; Mendoza, E.; Feagin, R. Response of vegetated dune–beach systems to storm conditions. *Coast. Eng.* **2016**, *109*, 53–62. [CrossRef]
49. De Vries, J.v.T.; Van Gent, M.; Walstra, D.; Reniers, A. Analysis of dune erosion processes in large-scale flume experiments. *Coast. Eng.* **2008**, *55*, 1028–1040. [CrossRef]

Journal of
Marine Science and Engineering

MDPI

Article

A Comparison of Beach Nourishment Methodology and Performance at Two Fringing Reef Beaches in Waikiki (Hawaii, USA) and Cadiz (SW Spain)

Juan J. Muñoz-Perez [1,*,†], Shari L. Gallop [2,†] and Luis J. Moreno [3,†]

1 Applied Physic Department, University of Cadiz, 11510 Puerto Real, Spain
2 School of Science and Environmental Research Institute, University of Waikato, 3110 Tauranga, New Zealand; shari.gallop@waikato.ac.nz
3 Department of Civil Engineering: Hydraulics, Energy and Environment, Technical University of Madrid, 28040 Madrid, Spain; luisjuan.moreno@upm.es
* Correspondence: juanjose.munoz@uca.es
† All authors contributed equally.

Received: 14 March 2020; Accepted: 4 April 2020; Published: 9 April 2020

check for updates

Abstract: Fringing reefs have significant impacts on beach dynamics, yet there is little research on how they should be considered in beach nourishment design, monitoring, and conservation works. Thus, the behavior and characteristics of nourishment projects at two reef protected beaches, Royal Hawaiian Beach (RHB) in Hawaii, USA, and Victoria Beach (VB) in Cadiz, Spain, are compared to provide transferable information for future nourishment projects and monitoring in fringing reef environments. The nourishment cost at RHB was nine times higher than VB. This is partly due to lower total volume and a more complex placement and spreading method at RHB, despite the much closer borrow site at RHB. There was a significant difference in post-nourishment monitoring frequency and assessment of accuracy. RHB elevation was monitored quarterly for 2.7 years at 30 m-spaced profiles, compared to 5 years of biannual surveys of 50 m-spacing at VB. An additional problem related to the presence of reefs at both RHB and VB was estimating the beach volume increase after nourishment, due to variable definitions of the 'beach' area and high alongshore variability in reef topography. At sites where non-native sediment is used, it is imperative to understand how wave and current energy changes due to reefs will influence nourishment longevity. Thus, differences in erosion and accretion mechanisms at both beaches have been detected, though are still little understood. Moreover, discrepancies in sediment porosity between the two sites (which should be surveyed in future nourishments) have been found, probably due to differences in the nourishment sand transportation and distribution methods. In summary, more dialogue is needed to explicitly consider the influence of fringing reefs on coastal processes and beach nourishment projects.

Keywords: beach nourishment; perched beaches; monitoring; cost; volume density; geologically controlled beach

---

1. Introduction

Beach nourishment is a key 'soft engineering' approach used worldwide to remediate coastal erosion, in contrast to 'hard coastal engineering' which involves construction of defenses such as breakwaters. Beach nourishment is used on beaches with a variety of sedimentary characteristics, from sand to gravel [1]. Such beach fills are generally designed using simplified mathematical models that assume the geologic setting has no influence on the beach morphology [2]. One such commonly used model is the equilibrium beach profile, which is meant to reflect the beach profile that would occur if the forces (dominantly waves and water levels) were held constant for a sufficiently long time [2,3]. However, there is increasing research showing that the geologic setting plays a dominant role in contemporary beach morphodynamics [4–7]. In fact, some studies have taken into account the former geologic setting to design multifunctional artificial reefs that serve several purposes [8].

One common type of geologically-controlled beach is situated landward of areas of hard bottom, such as rock or coral reefs. Reefs have large impacts on coastal hydrodynamics, sediment transport, and morphology, such as by influencing water level fluctuations [9], wave set-up and cross-reef currents [10], wave breaking, wave-induced flows, and wave attenuation over reef platforms [11–15], which affect the beach morphodynamics tidal range at multiple scales [16–19], even under hurricane conditions [20,21].

Despite clear evidence that beaches with reefs behave differently to their non-reef counterparts, there has been scant attention given to how reefs should be considered in beach nourishment projects. A recent study by Habel et al. [22] focused on the methodology of a beach nourishment at Royal Hawaiian Beach (RHB), Waikiki, which is fronted by a fringing coral reef. They also analyzed the subsequent monitoring. We share their interest in beaches that are fronted by reefs and find this a welcome opportunity to start a discussion of how reefs should be considered in beach nourishment projects. Therefore, the goal of this paper is to compare the method and performance of beach nourishments in fringing reef environments, performed at RHB and Victoria Beach (VB) in SW Spain, in order to provide transferable information for future projects with similar fringing reef environments. We compare nourishment costs and methods for nourishment performing, total nourishment volume, fine sediment fraction, and post-nourishment monitoring techniques between the two beaches.

2. Location Descriptions

RHB in Hawaii and VB in Cadiz, SW Spain, are both supported by calcareous reef platforms (Figures 1 and 2, respectively). Beaches on both of these coasts have experienced significant long-term erosion, which is problematic for the nationally important tourism industries that they support. Erosional trends are expected to accelerate with sea level rise [23], which is currently about 6 mm/year in Hawaii [24], and 2–3 mm/year in Cadiz. The costs of adaptation strategies (including beach nourishment) are already being considered [25].

RHB extends 520 m alongshore in a crescent shape between the Royal Hawaiian groin at the western end of the beach and the Kuhio Beach Ewa groin at the eastern end. This beach has been losing volume at a rate of 760 ± 450 m^3/year, consistent with the design rate of 1,070 m^3/year, that is, the value used for the nourishment project [22]. Victoria Beach is 3 km long and is divided into two zones due to the existence of a transversal fault to the shoreline. The northern zone (~1.2 km long) has a quasi-horizontal rocky platform, the surface of which aligns with spring low tide level. The dominant swell comes from the west, which generates dominant littoral drift towards the southeast, with an average loss value (i.e., the sand volume lost every year) of approximately 30,000 m^3/year [26].

Southerly swell at RHB generates, over the summer months, an average significant wave height and period of 0.8 m and 13.1 s, respectively. Erosion is usually experienced under shorter period winter waves at this beach [22]. On the other side, the western swell in VB generates a significant wave height of about 0.7 m (similar to HRB wave), but the average period is less, approximately 7 s [27]).

A summary of noteworthy morphological and sedimentological characteristics for the two beaches is shown in Table 1, and a visualization of these morphological concepts is shown in Figure 3. RHB is microtidal, whereas VB is mesotidal, with spring tidal ranges of up to 0.9 and 3.8 m, respectively. All elevations are relative to the lowest low-water level (LLWL) (+ 0.00 m). Elevation of the reef flat coincides approximately with this datum, and therefore the reef platform is emerged at the low tide level for VB and almost emerged for RHB. Another difference is related to the nature of sediment, which is calcareous in RHB, and almost completely silicic in VB (where the sand is 90–95% quartz and 5–10% bioclastic material). Both beaches have fine/medium sand, although the median grain size (D50) is slightly larger, at 0.34 mm for RHB, and 0.25 mm at VB.

A final key difference is that while both beaches have their toes supported by a reef platform, VB also has a fault scarp along its entire front and lacks the offshore channel which characterizes RHB. At RHB, this offshore channel is thought to be an important conduit for cross-shore sediment transport [22], a topic also mentioned by other researchers [28].

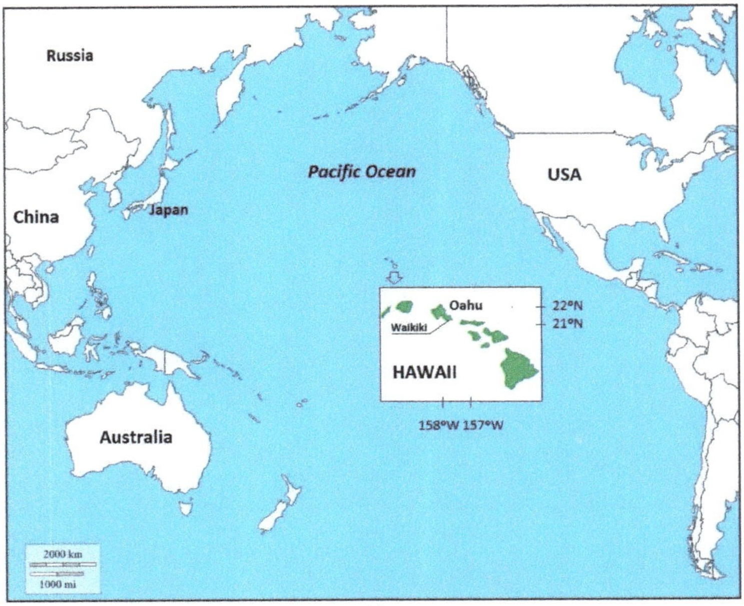

(a)

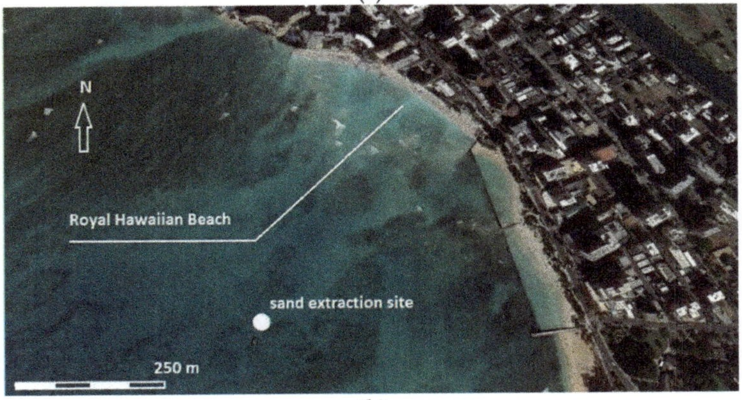

(b)

Figure 1. Location of Royal Hawaiian Beach (RHB) in Waikiki, Hawaii, USA (**a**); and an aerial view where the fringing reef can be observed and the sand extraction site is located (**b**) (image source: Google Earth 2018).

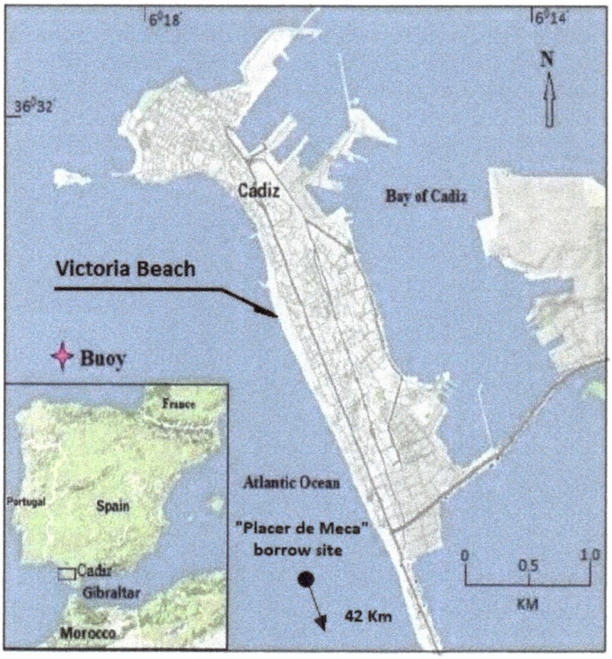

(a)

(b)

Figure 2. Location of Victoria Beach (VB) and indication of the borrow site (named Placer de Meca) in Cadiz, SW Spain, (**a**); and a ground view of the reef platform (**b**) (source: Muñoz-Perez, 2009).

Table 1. Comparison between Royal Hawaiian Beach (RHB) and Victoria Beach (VB) characteristics.

Beach	Slope (%)	D50 (mm)	Reef Width (m)	Tidal Range (m)	Berm Elevation (m)	Toe Elevation (m)	Latitude	Nature of Sediment
RHB	13.3	0.34	1000	0.9	+2.0	−1.0	21° N	Calcareous
VB	4.5	0.25	500	3.80	+4.0	+0.0	36° N	Silicic

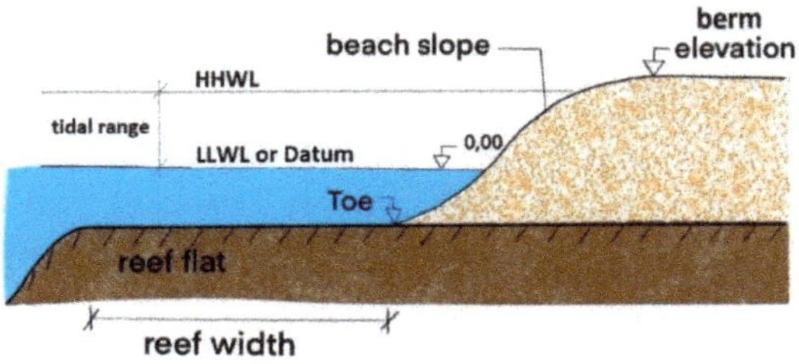

Figure 3. Sketch clarifying different morphological terms used in the text. LLWL refers to the lowest low water level.

3. Nourishment Methodology and Monitoring

Here, we compare and contrast the nourishment techniques and post-nourishment monitoring at RHB and VB.

3.1. Nourishment Method

There was a big difference in the method of placing, conveying and spreading the sand between RHB and VB. At RHB, 18,350 m^3 of sand was extracted from a source located approximately 0.6 km offshore of the nourishment site (Figure 1), using a pump that was suspended from a crawler crane stationed on a barge [22]. Sand was first pumped to an onshore dewatering basin adjacent to the placement location, before being distributed onto the beach using the truck haul method. This intermediate step was likely an important driver of the higher nourishment cost at RHB, discussed further in the next section.

In contrast, sand was pumped directly onto the dry beach at VB (i.e., the zone from high tide level to the landward edge). A volume of 260,000 m^3 was extracted from a borrow site called Placer de Meca, located 42 km (23 nm) from VB, close to the Cape of Trafalgar (Figure 2), using a trailing suction hopper dredger (TSHD). The dredger extracted the sand from a depth of ~20 m, and the complete process (mining, going to the beach, delivery, and returning to the borrow site) took about 8 h. Some pictures showing different aspects of the sediment pouring are shown in Figure 4.

3.2. Determination of Nourishment Volume

Methods used to quantify beach nourishment volumes to determine payment vary and can have significant implications for total nourishment costs. Although there was no specific indication in Habel et al. [22] of the measurement and payment guidelines used at RHB, they state that the initial post-nourishment survey, accomplished 10 days following the completion of sand placement, was used to determine the total nourished volume. During the first beach nourishments performed in Spain in the 1980s, payments to dredging companies were based on volume calculated by comparing pre- and post-nourishment beach profiles. A prominent dredging company at the time adduced that they were not to be held responsible for any possible erosion that may happen to nourished beaches caused by wave or tidal activity. The basis for this discrepancy was later discussed by Rullens et al. [29]. The final court decision was to their advantage, and since then payment has been made on hopper measurements taken before the sand sediment is pumped onto the beach. However, pre- and post-volume calculations are still undertaken, but only for monitoring purposes.

Concerns about the accuracy of the volumes measured aboard the ship are common. Readers interested in a detailed description of the methodology, and about an independent and portable meter system for dry weight control in hopper dredgers in order to distinguish between the dry sand weight and the sea water weight can consult [30].

Figure 4. Different aspects of the sediment pouring on Victoria Beach, showing (**a**) pipe assembly; (**b**) transport of the pipe to the sea; (**c**) aerial view of the pipe while pouring; and (**d**) pipe outlet onto the beach.

3.3. Post-Nourishment Monitoring

Post-nourishment monitoring is important to confirm the design life of the project and to assist in future planning. At RHB, beach elevation data were collected quarterly along cross-shore profiles for 2.7 years after nourishment. At VB elevation campaigns were performed biannually for 5 years post-nourishment. According to US Army Corps of Engineers (USACE) recommendation [31], monitoring beach nourishment requires large numbers of topo-bathymetric surveys (quarterly during the year following nourishment works, and biannually for at least one additional post-nourishment year).

At RHB, the shore-normal profiles were established at ~30 m intervals along the 520 m-long site, and extended either beyond the seaward toe of the fringing reef, or where fringing reef was absent to depths of 2–3 m. Profiles were surveyed by tracking a swimmer moving a rod-mounted prism across the beach, into nearshore waters, and over the fringing reef. Surveying was carried out randomly with respect to wave state and tidal cycle. The main sources of elevation error were identified and analyzed, including survey measurement error (by using established control points) and surface interpolation error (by conducting a comparison between random real and interpolated elevations). This is particularly noteworthy because only a handful of investigations have described errors related to typical beach bathymetric data sets and their influence on quantitative and qualitative interpretations of nearshore processes [32]. Grosskopf and Kraus [33] recommend a mean error of less than 11.5 m^3/m of beach volume to estimate sand volume with an accuracy comparable to the 10% to 20% contingencies associated with project designs. Following this line of reasoning, instrument error was not monitored at RHB, as the measurement device (Leica TC407 total station) had millimeter accuracy, although accuracy can diminish when this topographic procedure is applied to seaward zones, due to the inclination of the rod where the prism is mounted. A method to take into account and reduce this error was developed and described by [34].

On the other hand, at VB the sand was extended with a berm width of 100 m, and the berm elevation was about 1 m above the spring tidal level (+ 3.80 m over the datum). High resolution topographic levellings were carried out biannually during the following five years to account for sea-swell seasonality. Another important aspect of the post-monitoring survey method relates to the accuracy of the surveys. Surveying at RHB was undertaken regardless of the stage of the tidal cycle, due to the relatively low tidal range of 0.9 m (Table 1). H owever, at VB surveys were always performed at LLWL during spring tides. Surveys took two days, and thus the tidal range did not change appreciably during this monitoring. Cross-shore profiles were separated by 50 m alongshore and extended from the sea wall to the toe of the profile, even if submerged (the toe of the profile was always 1 m below LLWL). Bathymetries were considered unnecessary and would also likely have a large error due to the random roughness of the rock. Therefore, the maximum error detected for the levellings performed with a total station theodolite (according to the technique described in [35]) was <1 cm, and standard deviation was <0.5 cm.

In addition, VB is more gently sloping than RHB, at 4.5% compared to 13.3% (Table 1). This means that variations in the water level due to spring and neap tides could have a significant influence on volume measurements, as this can cause the sand level to vary by up to 0.67 m under negligible wave conditions at Cadiz [36].An additional difference in the post-nourishment monitoring was in the alongshore spacing of the cross-shore beach profiles. Some recommendations regarding beach profile spacing range from 300 m on long straight beaches with a 50% reduction close to structures or special points [31,37], to less than 150 m, according to the recommendation of the Coastal Engineering Manual [38]. Nevertheless, generally pre- and post-nourishment profile surveys are routinely collected at higher resolutions (with alongshore spacing of 60 m or less) to determine placement volumes accurately for payment purposes. Profiles at RHB were established at the relatively small spacing of 30 m, presumably to increase accuracy and decrease error associated with discretizing the domain. Although increasing the number of profiles measured decreases the error associated with discretizing the domain, it also increases the survey budget. Therefore, in order to find a compromise between a 5% allowable error of estimation and the available monitoring budget, the methodology developed in [35] was used to determine a 50 m profile spacing at VB [39,40]. Optimizing profile spacing is important for cost savings, however, while techniques exist for sandy beaches, it is not clear how reef variability influences the spacing required to capture key changes with sufficient alongshore variability.

4. Results and Discussion

4.1. Performance of Nourishment Works

The methods of nourishment material transport and distribution have significant implications for the resulting beach sediment characteristics. Habel et al. suggested that the truck haul method caused chemical compaction of nourished sands, which has important implications for slope stability [41] and ecological function [42]. In contrast, at VB sand porosity increased after the renourishments. This opposite behavior is probably due to the aforementioned differences in transport and distribution of nourishment material. While trucks transported dewatered sand at RHB, at VB there was a massive dumping of a mixture of sediment (20%) and water (80%) onto the backshore and foreshore from a trailing suction hopper dredge. According to Roman-Sierra [43] this procedure causes a significant increase in sand porosity, which subsequently decreases until it reaches its native value (about a year later), due to wave action driving spatial re-accommodation of the grains. Moreover, much of the volume loss after beach nourishment procedure can be due to a decrease in porosity. A detailed analysis of the impacts on porosity of the sand transport and nourishment method, as well as an accurate and novel application of in situ measurements of the porosity of beach sand using a high-quality nuclear densimeter gauge, are also described in [43]. We suggest that measuring sediment porosity in future nourishments is advisable to determine if the transportation and distribution method significantly affects porosity, and hence beach slope stability and sand volume. The RHB nourishment project plans to maintain beach width for 20 years, with the initial replenishment plus an additional second phase after 10 years. This compares to a design life of 25 years for VB, although a small replenishment (about 25% of the original nourishment volume) was undertaken after 12 years, due to high social demand.

4.2. Nourishment Volumes

In reference to the measurements of placed sand volume conducted daily by contractors in RHB, a total estimated volume of 17,551 m^3 was established. However, using the meritorious interpolation error methodology presented by Habel et al. [22], a beach volume increase of 12,700 ± 3700 m^3 and a system volume increase of 13,700 ± 6300 m^3 were confirmed. Revisiting some explanations about "beach" and "system" concepts are noteworthy. According to the aforementioned authors, the beach area extends a nominal 20 m seaward of the initial post-nourishment beach toe, while the system area extends an additional 150 m seaward. Regrettably, despite the laudability of the method, estimation of error becomes 30% at the "beach" zone and rises to 46% when the "system" zone is considered.

The error in sand volume estimation was somewhat lower in VB, probably due to the larger volume poured; 260,000 m^3 was measured into the hopper of the dredger vs. 283,000 m^3 surveyed on the beach, due to the increase of the porosity [43]. Another source of error was probably the existence of a problem at VB similar to that of RHB: the great variability of the levellings over the fringing reef. Ultimately, it was decided to exclude the reef from the surveying zone due to the lack of credibility.

Related to the seasonal erosion/accretion cycle, an unexpected volume gain occurred at RHB following increases in incident wave energy flux above 10,000 kg m^3/s^3. As Habel et al. [22] stated, it would be interesting to obtain elevation data from regions further offshore of their present study area which could confirm the existence of a proximal sediment source/sink. That would permit finding the relationship between the pattern of erosion/accretion and cross-shore sediment transport. On the contrary, significant gaining was never detected at VB. An average erosion value of 10,000 m^3/year, which resulted in an irreversible loss [26], was obtained. However, this erosion rate was much smaller than for non-reef protected profiles subjected to similar wave energy on the same beach (30,000 m^3/year).

4.3. Nourishment Costs

There was a large difference in nourishment costs between the two beaches. At RHB, the cost was 2.9 million USD for recovery, dewatering, and emplacement of 18,350 m^3 along a 520 m segment of coastline. This investment represents an overall average price of \$158 USD/m^3. It should be noted that the cost of removal of two dilapidated sandbag groin structures has not been subtracted, because is not available. In comparison, the cost of nourishment at VB, where sand was dredged locally from the seabed, ranged from 12 to 15 €/m^3 (13–17 USD/m^3 at the current rate of exchange) for the various nourishments, excluding taxes. Moreover, distance from the borrow site in Cadiz (close to the Cape of Trafalgar) was 37–46 km, compared to just 0.6 km at RHB. However, costs of mobilization and demobilization of equipment in Hawaii account for much of the project costs, due to how remote the islands are. It is also true that supplied volumes were far superior in VB, with over 200,000 m^3 at all nourishment actions, thus allowing for a lower unit cost of sand. Further details about beach nourishments carried out from 1999 to 2010, storm climate, and sea level seasonal cycle in the Gulf of Cadiz, can be found in [44–46] respectively.

4.4. Fine Sediments

Consideration of the fine sediment fraction in nourishment material is important, not only because of its contribution to elevated turbidity, but also because fines almost immediately are transported offshore, leading to immediate loss of sediment volume on the beach [22]. At RHB, Habel et al. stated that no fines were present in the mined sand field prior to dredging and transport. Thus, the existence of a fine sediment plume observed during the nourishment procedure was probably due to carbonate dissolution [22]. This point is interesting, because the fine-sediment percentage was also negligible at the samples taken from Trafalgar (the borrow site used for VB), but nevertheless, there was always dredging-induced turbidity which decreased with time, reaching natural conditions approximately 9 min after the operations ceased [47]. Since Trafalgar sand is siliceous (and therefore the possibility of calcareous solution must be discarded), we should consider the plausibility of losses of the fine-sediment fraction when a Van Veen grab is used to sample the sediment, instead of another procedure which guarantees good fine material retention and undisturbed samples.

4.5. Considering Reefs in Nourishment Design

No specific methodology was applied to consider the reef in the design process of RHB fill. Habel et al. [22] used a similar sand, and beach slope was predicted to reach an equilibrium profile similar to the pre-nourished beach, which presumably was in equilibrium with the hydrodynamic conditions created by the reef. For a case where borrowed and native sands are not identical, wave energy dissipation over the reef (due to wave breaking or bottom friction) should be taken into account [48]. These fringing reef beaches are so usual that some numerical models (e.g., SBEACH or SMC) have been modified, trying to allow calculation of the profile [49,50]. Moreover, there are a lot of characteristics typical of this kind of beach. It is often assumed that perched beaches are more stable in time than non-perched beaches [51–53], through wave energy dissipation (e.g., [11,13]) and reducing rates of longshore sediment transport [54]. However, Gallop et al. [6] found that depending on the reef topography, reefs can generate current jets that can increase sediment transport on and off the beachface over hourly timescales. Moreover, over seasonal timescales, reefs can create highly dynamic seasonal changes in beach morphology [28].

Annual and interannual changes in beach morphology are also important on reef beaches [55]. The temporal and spatial variability of these oscillations can be large (e.g., [7,55]), but are difficult to measure, let alone incorporate into nourishment design. Leaving aside some easy-to-apply remedies, such as the modification on the A parameter of the Dean's formula [56], there are few engineering tools developed to take into account the idiosyncrasy of reef fringing beaches in nourishment design. Understanding the influence of these fringing reefs on beach behavior is imperative for coastal management, not only from the scientific point of view, but also the economic.

Finally, although not the focus of this paper, the potential impacts of sea level rise due to climate change on nourishment projects in fringing reef environments should be highlighted as a future line of research. Some research has already been undertaken on this topic at study sites, such as simulation of groundwater inundation in Honolulu [57], assessing vulnerability of low-lying areas in Maui [58] or in Oahu [59], or criticizing elevation levels for flooding due to sea level rise in Hawaii [60]. Moreover, mapping inundation probability has already been performed in some zones of the Gulf of Cadiz, Spain [61]. Due to concerns about the long-term performance of beach nourishment with sea level rise, different types of adaptation measures are been considered, such as seawalls or breakwaters, accommodating sea level rise by raising buildings, and even managed retreat. However, plans for these two sites have not yet been decided.

5. Conclusions

Characteristics of two reef protected beaches, RHB in Hawaii, USA, and VB in Cádiz, Spain, were compared to provide transferable information for future projects in areas with similar fringing reef environments. Results show that the complexity of the spreading method is a key determinant of overall cost, as well as total nourishment volume, which impacts on the cost per unit volume. This comparison has shown that more research is needed for determining optimal profile spacing, because while techniques exist for sandy beaches, it is not clear how reef variability influences the spacing required to capture key changes. It is also unclear how to include variable reef topography when estimating total nourishment volumes to determine nourishment volumes, and hence costs. Moreover, erosion and accretion mechanisms, and their temporal and spatial variability on reef beaches, are little understood, and hence difficult to incorporate into nourishment design. Sediment porosity surveying would be advisable in future nourishments to determine how porosity (and hence beach slope stability and sand volume) is affected by the transportation and distribution method. Publishing of studies of nourishment projects on reef beaches is rare. More dialogue and sharing of experiences are key to improving coastal zone management, optimizing nourishment projects, and preserving our precious beaches.

Author Contributions: All authors contributed equally. All authors have read and agreed to the published version of the manuscript.

Funding: This research received no external funding.

Conflicts of Interest: The authors declare no conflict of interest.

References

1. Hanson, H.; Brampton, A.; Capobianco, M.; Dette, H.H.; Hamm, L.; Laustrup, C.; Lechuga, A.; Spanhoff, R. Beach nourishment projects, practices and objectives—A European overview. *Coast. Eng.* **2002**, *47*, 81–111. [CrossRef]
2. Dean, R.G. Equilibrium beach profiles: Characteristics and applications. *J. Coast. Res.* **1991**, *7*, 53–84.
3. Pilkey, O.H.; Young, R.S.; Riggs, S.R.; Smith, A.W.S.; Wu, H.; Pilkey, W.D. The concept of shoreface profile of equilibrium: A critical review. *J. Coast. Res.* **1993**, *9*, 255–278.
4. Thieler, R.E.; Brill, A.L.; Cleary, W.J.; Hobbs, C.H.; Gammisch, R.A. Geology of the Wrightsville Beach, North Carolina shoreface: Implications for the concept of shoreface profile of equilibrium. *Mar. Geol.* **1995**, *126*, 271–287. [CrossRef]
5. Contreras de Villar, A.; Gómez-Pina, G.; Muñoz-Pérez, J.J.; Contreras, F.; López-García, P.; Ruiz-Ortiz, V. New design parameters for biparabolic beach profiles (SW Cadiz, Spain). *Rev. Construcción* **2019**, *18*, 432–444. [CrossRef]
6. Gallop, S.L.; Bosserelle, C.; Pattiaratchi, C.; Eliot, I. Rock topography causes spatial variation in the wave, current and beach response to sea breeze activity. *Mar. Geol.* **2011**, *290*, 29–40. [CrossRef]
7. Gallop, S.L.; Bosserelle, C.; Haigh, I.D.; Wadey, M.P.; Pattiaratchi, C.; Eliot, I. The impact of temperate reefs on 34 years of shoreline and vegetation line stability at Yanchep, southwestern Australia and implications for coastal setback. *Mar. Geol.* **2015**, *369*, 224–232. [CrossRef]
8. Antunes, J.S.; Neves, G.; ten Voorde, M. Designing a multifunctional artificial reef: Studies on the influence of parameters with most influence in the vertical plane. *J. Coast. Conserv.* **2011**, *15*, 151–157. [CrossRef]
9. Karunarathana, H.; Tanimoto, K. Numerical experiments on low-frequency fluctuations on a submerged coast reef. *Coast. Eng.* **1995**, *26*, 271–289. [CrossRef]
10. Symonds, G.; Black, K.P.; Young, I.R. Wave-driven flow over shallow reefs. *J. Geophys. Res.* **1995**, *100*, 2639–2648. [CrossRef]
11. Gourlay, M.R. Wave transformation on a coral reef. *Coast. Eng.* **1994**, *23*, 17–42. [CrossRef]
12. Coronado, C.; Candela, J.; Iglesias-Prieto, R.; Sheinbaum, J.; Lopez, M.; Ocampo-Torres, F.J. On the circulation in the Puerto Morelos fringing reef lagoon. *Coral Reefs* **2007**, *26*, 149–163. [CrossRef]
13. Lowe, R.J.; Hart, C.; Pattiaratchi, C.B. Morphological constraints to wave-driven circulation in coastal reef–lagoon systems: A numerical study. *J. Geophys. Res.* **2010**, *115*. [CrossRef]
14. Hoeke, R.; Storlazzi, C.; Ridd, P. Hydrodynamics of a bathymetrically complex fringing coral reef embayment: Wave climate, in situ observations, and wave predictions. *J. Geophys. Res.* **2011**, *116*, 1–19. [CrossRef]
15. Mendonça, A.; Fortes, C.; Capitão, R.; Neves, M.G.; Moura, T.; Antunes do Carmo, J.S. Wave hydrodynamics around a multi-functional artificial reef at Leirosa. *J. Coast. Conserv.* **2012**, *16*, 543–553. [CrossRef]
16. Sheppard, C.; Dixon, D.J.; Gourlay, M.; Sheppard, A.; Payet, R. Coral mortality increases wave energy reaching shores protected by reef flats: Examples from the Seychelles. *Estuar. Coast. Shelf Sci.* **2005**, *64*, 223–234. [CrossRef]

17. Gallop, S.L.; Bosserell, C.; Eliot, I.; Pattiaratchi, C.B. The influence of limestone reefs on storm erosion and recovery of a perched beach. *Cont. Shelf Res.* **2012**, *47*, 16–27. [CrossRef]
18. Ruiz, A.; Mariño-Tapia, I.; Enriquez, C.; Silva, R.; González-Leija, M. The role of fringing coral reefs on beach morphodynamics. *Geomorphology* **2013**, *198*, 69–83.
19. Rizzo, A.; Aucelli, P.; Gracia, F.J.; Anfuso, G. A novelty coastal susceptibility assessment method: Application to Valdelagrana area (SW Spain). *J. Coast. Conserv.* **2017**, *5*, 51. [CrossRef]
20. Mariño, I.; Enriquez, C.; Silva, R.; Mendoza, E.; Escalante, E.; Ruiz, F. Comparative morphodynamics between exposed and reef protected beaches under hurricane conditions. *Coast. Eng. Proc.* **2014**, *1*, 55. [CrossRef]
21. Martell, R.; Silva, R.; Mendoza, E.G.; Muñoz, J.J.; Cerdeira, S.; Laiz, I. Spectral bimodality of waves produced by hurricanes in the Caribbean coastal zone of Mexico. *Cienc. Mar.* **2018**, *44*, 33–48. [CrossRef]
22. Habel, S.; Fletcher, C.H.; Barbee, M.; Anderson, T.R. The influence of seasonal patterns on a beach nourishment project in a complex reef environment. *Coast. Eng.* **2016**, *116*, 67–76. [CrossRef]
23. Anderson, T.R.; Fletcher, C.H.; Barbee, M.M.; Frazer, L.N.; Romine, L.N. Doubling of coastal erosion under rising sea level by mid-century in Hawaii. *Nat. Hazards* **2015**, *78*, 75–103. [CrossRef]
24. Hawaii's Sea Level is Rising. Available online: sealevelrise.org/states/hawaii/ (accessed on 9 March 2020).
25. Climate Change in Spain. Available online: www.climatechangepost.com/spain (accessed on 9 March 2020).
26. Muñoz-Perez, J.J.; Medina, R. Comparison of long-, medium and short-term variations of beach profiles with and without submerged geological control. *Coast. Eng.* **2010**, *57*, 241–251. [CrossRef]
27. Puertos del Estado (Spain). Available online: www.puertos.es (accessed on 9 March 2020).
28. Gallop, S.L.; Bosserelle, C.; Eliot, I.; Pattiaratchi, C.B. The influence of coastal reefs on spatial variability in seasonal sand fluxes. *Mar. Geol.* **2013**, *244*, 132–143. [CrossRef]
29. Rullens, R.C.; d'Angremond, K.; Ottevanger, G. Tonnes dry solids reviewed. In *Dredging'94 Proceedings of the 2nd International Conference on Dredging and Dredged Material Placement*; ASCE: New York, NY, USA, 1994; pp. 673–682.
30. Muñoz-Perez, J.J.; Gutierrez, J.M.; Moreno, J.; Español, L.; Moreno, L.; Bernabeu, A. Portable meter system for dry weight control in dredging hoppers. *J. Waterw. Port Coast. ASCE* **2003**, *129*, 79–85. [CrossRef]
31. Wise, R.A. Recommended base-level physical monitoring of beach fills. In *CETN II-35. U.S. Army Corps of Engineers*; Coastal Engineering Research Center: Vicksburg, MS, USA, 1995.
32. Plant, N.G.; Holland, K.T.; Puleo, J.A. Analysis of the scale of errors in nearshore bathymetric data. *Mar. Geol.* **2002**, *191*, 71–86. [CrossRef]
33. Grosskopf, W.G.; Kraus, N.C. Guidelines for surveying beach nourishment projects. In *Technical Note*; CETN II-31; U.S. Army Engineering Water Experiment Station: Vicksburg, MS, USA, 1993; p. 12.
34. Serra, J.; Medina, J.R. Beach monitoring program of Valencia (Spain). In Proceedings of the 25th International Conference on Coastal Engineering Proceedings, Orlando, FL, USA, 2–6 September 1996; pp. 2871–2883.
35. Muñoz-Perez, J.J.; Khan-Mozahedy, A.B.M.; Neves, M.G.; Tejedor, B.; Gomez-Pina, G.; Campo, J.M.; Negro, V. Sinking of concrete modules into a sandy seabed: A case study. *Coast. Eng.* **2015**, *99*, 26–37. [CrossRef]
36. Muñoz-Perez, J.J.; Medina, R. Profile changes due to a fortnightly tidal cycle. In *Proceedings of the 28th ICCE*; ASCE: Sydney, Australia, 2002; pp. 3063–3075.
37. Browder, A.E.; Dean, R.G. Monitoring and comparison to predictive models of the Perdido Key beach nourishment project, Fl., USA. *Coast. Eng.* **2000**, *39*, 173–191. [CrossRef]
38. USACE (US Army Corps of Engineers). *Coastal Engineering Manual*; EM 1110-2-1100; USACE: Washington, DC, USA, 2002.
39. Muñoz-Perez, J.J.; Payo, A.; Roman-Sierra, J.; Navarro, M.; Moreno, L. Optimization of beach profile spacing: An applicable tool for coastal monitoring. *Sci. Mar.* **2012**, *76*, 791–798. [CrossRef]
40. Muñoz-Perez, J.J.; Navarro, M.; Roman, J.; Tejedor, B.; Gomez, G. Long-term evolution of a transgressive migrating dune using reconstruction of the EOF method. *Geomorphology* **2009**, *112*, 167–177. [CrossRef]
41. Croize, D.; Bjørlykke, K.; Jahren, J.; Renard, F. Experimental mechanical and chemical compaction of carbonate sand. *J. Geophys. Res. Solid Earth* **2010**, *115*. [CrossRef]
42. USACE (US Army Corps of Engineers). Beach erosion control projects for Palm Beach County, FL, USA. In *General Design Memorandum with Environmental Impact Statement*; USACE: Washington, DC, USA, 1987.
43. Roman-Sierra, J.; Muñoz-Perez, J.J.; Navarro-Pons, M. Beach nourishment effects on sand porosity variability. *Coast. Eng.* **2014**, *83*, 221–232. [CrossRef]

44. Muñoz-Perez, J.J.; Roman, J.; Navarro-Pons, M.; da Graça, M.; del Campo, J.M. Comments on Confirmation of beach accretion by grain-size trend analysis: Camposoto beach, SW Spain by Poizot, E., et al. *Geo-Mar. Lett.* **2014**, *34*, 75–78.
45. Plomaritis, T.A.; Benavente, J.; Laiz, I.; del Río, L. Variability in storm climate along the Gulf of Cadiz: The role of large scale atmospheric forcing and implications to coastal hazards. *Clim. Dyn.* **2015**, *45*, 2499–2514. [CrossRef]
46. Laiz, I.; Tejedor, B.; Gomez-Enri, J.; Aboitiz, A.; Villares, P. Contributions to the sea level seasonal cycle within the Gulf of Cadiz (Southwestern Iberian Peninsula). *J. Mar. Syst.* **2016**, *159*, 55–66. [CrossRef]
47. Roman-Sierra, J.; Navarro, M.; Muñoz-Perez, J.J.; Gomez, G. Turbidity and other effects resulting from Trafalgar sandbank dredging and Palmar beach nourishment. *J. Waterw. Port Coast. ASCE* **2011**, *137*, 332–343. [CrossRef]
48. Massel, S.R.; Gourlay, M.R. On the modelling of wave breaking and set-up on coral reefs. *Coast. Eng.* **2000**, *39*, 1–27. [CrossRef]
49. Larson, M.; Kraus, N. Representation of non-erodible (hard) bottoms in beach profile change modeling. *J. Coast. Res.* **2000**, *16*, 1–14.
50. Gonzalez, M.; Medina, R.; Gonzalez, J.; Osorio, A.; Mendez, F.J.; Garcia, E. An integrated coastal modeling system for analyzing beach processes and beach restoration projects, SMC. *Comput. Geosci.* **2007**, *33*, 916–931. [CrossRef]
51. Bernabeu, A.M.; Muñoz-Perez, J.J.; Medina, R. Influence of a rocky platform in the profile morphology: Victoria Beach, Cádiz (Spain). *Cienc. Mar.* **2002**, *28*, 181–192. [CrossRef]
52. Vousdoukas, M.I.; Velegrakis, A.F.; Plomaritis, T.A. Beachrock occurrence, characteristics, formation mechanisms and impact. *Earth Sci. Rev.* **2007**, *85*, 23–46. [CrossRef]
53. Moreno, L.; Negro, V.; Garrote, L.; Muñoz-Perez, J.J.; Lopez, J.S.; Esteban, M.D. An Engineering Method for the Preliminary Functional Design of Perched Beaches. Theoretical Approach. *J. Coast. Res.* **2018**, *85*, 1261–1265. [CrossRef]
54. Eversole, D.; Fletcher, C.H. Longshore sediment transport rates on a reef-fronted beach: Field data and empirical models Kaanapali Beach, Hawaii. *J. Coast. Res.* **2003**, *19*, 649–663.
55. Norcross, Z.M.; Fletcher, C.H.; Merrifield, M. Annual and interannual changes on a reef-fringed pocket beach: Kailua Bay, Hawaii. *Mar. Geol.* **2002**, *190*, 553–580. [CrossRef]
56. Muñoz-Perez, J.J.; Tejedor, L.; Medina, R. Equilibrium beach profile model for reef-protected beaches. *J. Coast. Res.* **1999**, *15*, 950–957.
57. Habel, S.; Fletcher, C.H.; Rotzoll, K.; El-Kadi, A.I. Development of a model to simulate groundwater inundation induced by sea-level rise and high tides in Honolulu, Hawaii. *Water Res.* **2017**, *114*, 122–134. [CrossRef]
58. Cooper, H.M.; Chen, Q.; Fletcher, C.H.; Barbee, M.M. Assessing vulnerability due to sea-level rise in Maui, Hawaii using LiDAR remote sensing and GIS. *Clim. Chang.* **2013**, *116*, 547–563. [CrossRef]
59. Onat, Y.; Francis, O.P.; Kim, K. Vulnerability assessment and adaptation to sea level rise in high-wave environments: A case study on Oahu, Hawaii. *Ocean Coast. Manag.* **2018**, *157*, 147–159. [CrossRef]
60. Kane, H.H.; Fletcher, C.H.; Frazer, L.N.; Barbee, M.M. Critical elevation levels for flooding due to sea-level rise in Hawaii. *Reg. Environ. Chang.* **2015**, *15*, 1679–1687. [CrossRef]
61. Fraile-Jurado, P.; Álvarez-Francoso, J.I.; Guisado-Pintado, E.; Sánchez-Carnero, N.; Ojeda-Zújar, J.; Leatherman, S.P. Mapping inundation probability due to increasing sea level rise along El Puerto de Santa María (SW Spain). *Nat. Hazards* **2017**, *87*, 581–598. [CrossRef]

Journal of
Marine Science and Engineering

MDPI

Article

Micro Sand Engine Beach Stabilization Strategy at Puerto Morelos, Mexico

Mireille Escudero, Edgar Mendoza * and Rodolfo Silva

Institute of Engineering, National Autonomous University of Mexico, 04510 CdMx, Mexico; mescuderoc@iingen.unam.mx (M.E.); rsilvac@iingen.unam.mx (R.S.)
* Correspondence: emendozab@iingen.unam.mx; Tel.: +52-5556-233600 (ext. 8957)

Received: 27 February 2020; Accepted: 31 March 2020; Published: 2 April 2020

check for updates

Abstract: In the last decade, innovative beach nourishment strategies have been developed, driven by the increased worldwide interest in environmentally friendly coastal protection measures. In this context, the massive nourishment project of the Netherlands, known as Sand Engine, begun in 2011, has been hailed as a successful means of beach protection. Continuous monitoring, field campaigns, and numerical modeling have shown that the great volume of sand deployed is gradually transported by the waves and currents along the coastline, avoiding the need for repeated invasive, small scale beach replenishments. A very small, bell-shaped Sand Engine was designed to protect the beachfront at a tourist resort near Puerto Morelos, Mexico. To estimate the morphological response of the beach and the functioning of the micro Sand Engine as a sand reservoir, XBeach numerical modelling was applied to the project. The micro Sand Engine is seen to be a sustainable and eco-friendly coastal protection measure, especially applicable when a large nourishment project is not viable. Maintenance work for the nourishment is cost and time effective, and any negative impacts to sensitive ecosystems nearby can be detected and controlled quickly.

Keywords: beach nourishment; sand engine; XBeach; shore protection; beach restoration

1. Introduction

International interest in environmentally friendly strategies to protect beaches and dunes against erosion has led to an increasing preference for soft methods over hard engineering [1–4]. Of these soft methods, artificial beach nourishment has been recognized as particularly effective in curbing erosion, as it does not cause harmful effects to nearby areas [5].

Traditional beach nourishments, where the sand is placed directly ono the dry beach and dunes, is being superseded by large-scale shoreface nourishment, where the sand is placed between the low water line and the dry beach; it may be less expensive and provide recreational areas more quickly than the earlier mode.

Numerical models have been developed to better understand and optimize shoreface beach nourishment projects. Variations within the one-line shoreline models include that presented by [6], which calculates the shoreline evolution following a beach nourishment project, and considers the effect of using sediment with different characteristics than the native sediment; and the application of the numerical model GENESIS to the long-term simulation of shoreline change following a beach nourishment project in China [7]. Additionally, an analytical model and a One Line model were developed to compare beach performance with measured data in three beach projects in Florida [8]. A wave-sediment transport numerical model based on the higher order Boussinesq equations was developed by [9], which can calculate both cross-shore and planform morphology evolution. For the evaluation of shoreface nourishments specifically, [10] examined the development of a shoreface nourishment off Sylt Island, using a numerical model to simulate the hydro and local sediment transport processes, and the validity of XBeach numerical model was presented to predict the erosion of 19 shoreface nourishments at different sections of the Dutch coast [11]. Cross-shore transport (for shore-normal waves) governed the first-year erosion rates of the beach nourishment and alongshore transport only contributed about 15% to 40%, most erosion being produced under energetic wave conditions, and mild to moderate waves propagating without breaking over the nourishment [11].

In [12,13], high energy events, especially the first storm, sediment grain size, project extent, and longshore transport gradients have been reported as important factors in the first-year response of shoreface beach nourishments in a micro-tidal environment on the Florida Gulf Coast.

Improvements in alongshore sediment transport physics and in the knowledge on mechanical properties of sediment and placement, besides wave climatology forcing during the project, would lead to better prediction of project evolution [14]. The behavior of shoreface nourishment is not well understood [11] and it may be less effective if large beach nourishments are carried out in places where the possible beach response has not been previously well documented [14]. That is the case for many sand beaches in developing countries.

Modern coastal engineering requires sustainable coastal zone development that is compatible with the natural environment [15,16], thus Building with Nature (BwN) approaches are increasingly sought. In 1990, the Dutch government stated that beach nourishment was the preferred option in adapting to climate change. In consequence, an innovative large scale "sandscaping" beach nourishment project was begun in the Netherlands, using the so-called 'Sand Motor', 'Zandmotor', or 'Sand Engine' strategy [17–20]. An enormous volume of sand, approximately 20 Mm^3 (~10,000 m^3/m), was put onto the beach face between Ter Heijde and Kijkduin, in South Holland, from the low water mark to a distance of several kilometers off the coast, 20 m deep. The natural re-distribution of sediments to the nearby beaches and dunes is taking place through the action of wind, waves, and currents over several years [17]. The main objectives of this innovative BwN solution [17,19,21] were the creation of recreational areas that were environmentally safe. The project uses the local hydrodynamics in order to induce sedimentation through natural beach processes, rather than mechanical means, which can produce unfavorably steep beach slopes or bury marine organisms in the sand. Several Sand Engine designs were discussed before the pilot project was untaken in 2011, such as Full beach face nourishment, Peninsula or bell-shaped nourishment and the Hook configuration, which was eventually selected, Figure 1. This design was recommended for South Holland in the EIA (Environmental Impact Assessment, 2010), based on costs, safety, environmental and recreational criteria [17].

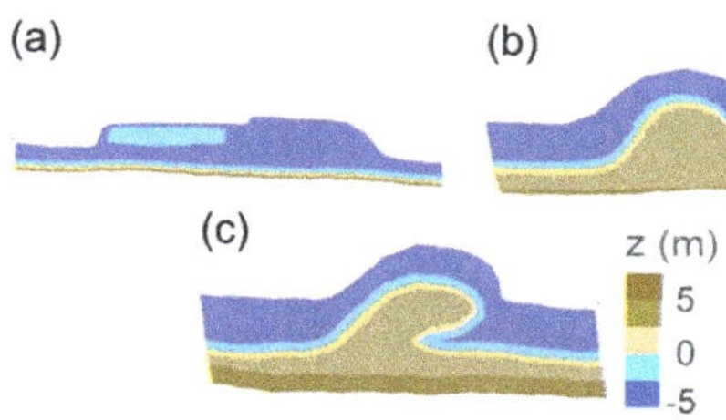

Figure 1. Three types of Sand Engine design (after [15]: (**a**) beach face nourishment; (**b**) bell-shaped nourishment; (**c**) hook-shaped nourishment.

Systematic field monitoring combined with numerical modelling, on a monthly and yearly basis, allows the processes and mechanisms that govern the dynamic of the Sand Engine to be forecasted, in the short and long term. The computation of the Delft3D model showed similar morphological behaviors for the three alternative designs described above, in the long term (20–50 years), with a nourishment lifetime of some twenty years [17]. The longshore sediment transport rate was the key parameter in defining how fast the Sand Engine shape would be eroded and the sediment be spread along the beach [22]. The numerical results show the greatest changes occur in the six months after construction, a moderate evolution in the following 12 months and asymptotic stabilization after the first year, as the slope and the new coastline reached an almost stable shape [22,23].

The results of the Defl3D model showed wave forcing to be the most important process in the initial morphological changes induced by the Sand Engine (accounting in the first year for 75% of the total volume lost) [24]. A linear beach face evolution was found; the duration of a storm event had greater effect than the magnitude of the maximum waves during a storm. This suggests that less energetic wave episodes, with a high probability of occurrence, were also important in the Sand Engine response in the first year [24,25].

The tidal range was found to be the second most important factor contributing to the morphological behavior of the Sand Engine; 17% of volumetric sediment loss was due to this, while less than 5 % of the total erosion was caused by surge, wind, and horizontal tidal forcing [24]. The combination of energetic wave conditions, strong winds, and high storm surge levels can lead to high sediment transport rates and therefore intense erosion [26]. The majority (~70%–72%) of the sediment lost from the Sand Engine was found to accrete in adjacent coastal sections and dunes, and this alongshore spread was strongly related to incident waves [23,24,27]. On the other hand, small wave heights produced mainly cross-shore sediment transport [23].

Similar estimations were also obtained from the application of other numerical models, such as the MIKE21FM shoreline model [28].

Several field monitoring techniques were used to validate the numerical results: (1) aerial photographs were used for observations [19], (2) remote sensing methods and in-situ measurements of the spatial and temporal variability of the Sand Engine [29], (3) field campaigns to measure water levels inside the hook-shaped lagoon [30], (4) a six-week campaign in the autumn of 2014 monitored the spatial variations in aeolian sediment transport [31], and (5) new methods of automated shoreline detection from satellite images [32].

In this paper, experiences from the Sand Engine project were applied for a beach in Mexico with more than ten years of continuous monitoring. This nourishment aims to improve beach conditions by natural processes. This beach seems to be promising for the long-term sediment dispersal of a micro Sand Engine project. A natural accumulation of sand very close to the breakwater of the marina has been observed in the past, which can be used as a sand budget for periodic reloading of the Sand Engine.

2. Study Area

The study area is near Puerto Morelos, in the NE of the Yucatan peninsula, in the state of Quintana Roo, on the Mexican Caribbean, 30 km south of the famous tourist destination of Cancun (Figure 2). The coastal stretch of the study area is the beach front of the Now Jade Riviera Cancun Resort, which is around 400 m in length (see Figure 2c,e). To the north, the area is limited by the Puerto Morelos marina. A considerable mangrove extension is the main land cover behind the shoreline.

The climate of the region is tropical, with two main seasons, winter and summer, in terms of wind patterns and air temperature. Winter (from November to March or April) is characterized by the passages of cold fronts, locally known as 'Nortes', with a wind direction from the northeast from October to February, though with the presence of winds from the north and the southeast following the cold fronts. In summer, the frequent storms and tropical cyclones define the patterns [33]. 90% of the records at buoy 42056 (NDBC-NOAA) show a relatively small significant wave height (<1 m) and small wave periods (<8 s), and thus the predominance of locally generated seas [34]. The fine sand is the results of this wave climate, but under hurricane conditions the beaches are very vulnerable to dramatic erosion processes. The beach is composed of medium carbonate sand of biogenic origin, with a mean sediment size of ~0.258 mm.

The beach is microtidal, predominantly semi-diurnal, with a mean tidal range of ~0.17 m, with spring and neap tidal ranges of 0.32 and 0.07 m, respectively [33]. Under extreme storm conditions, the storm surge is considered the most important threat. During hurricane Wilma, in October 2005 it reached 2.5 m at the beach (0.5 m in deep water) [34].

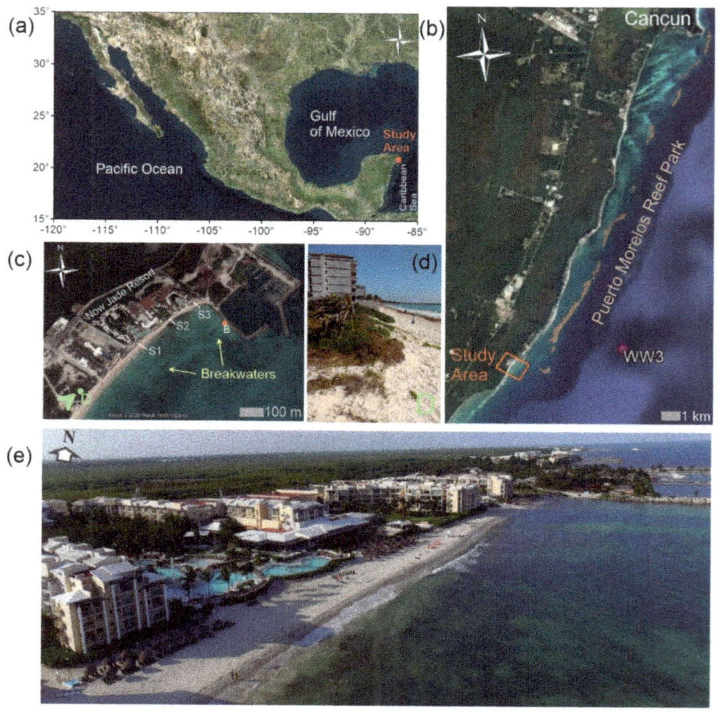

Figure 2. The study area: (**a**) general location; (**b**) Coast at Puerto Morelos; (**c**) aerial view of the Now Jade resort; (**d**) existing dunes, 250 m south of the study area, taken 26 April, 2017; (**e**) aerial view, taken 12 August 2019.

3. Data and Methods

3.1. Beach Evolution from Satellite Images and Fied Data

Ten satellite images taken from Google Earth (0.5-m resolution) were used to analyze the evolution of the hotel construction and the dry beach changes in the study area, in the period from 2004 to 2017. Topographic surveys were also carried out on 30 October 2012 and 26 April 2017 using a LEICA 3D scanner with 1–2 mm of precision, for the analysis of the beach evolution. Cross-shore profiles started at the limit of the hotel and extended to the submerged beach at a depth of 0.3 m. Nearshore bathymetric data were acquired using a double-frequency echosounder and complemented with 4 m resolution bathymetric data available from the National Biodiversity information system (http://www.conabio.gob.mx/informacion/gis).

3.2. Offshore Wave Data and Numerical Model Description

3.2.1. Offshore Wave Data

Hindcast data from the WAVEWATCH III model (WW3) provided the wave database for 1 February 2005 to 1 June 2019. Times series of wave data, composed of sea states at 3 h intervals, were supplied at 60 m depth, 4.5 km offshore, with coordinates 20.8333° N 86.8333° W (see Figure 2b). A storm was defined as an event with a significant wave height exceeding 1.6 m [35].

3.2.2. Numerical Model Description

The XBeach [36] numerical model was used to estimate the wave and current fields as well as the morphological evolution of the nourishment for six scenarios. The wave conditions were modelled using a Jonswap wave-spectrum with 30-h simulations, when morphological convergence was found.

Two sets of rectangular meshes were constructed to incorporate the bathymetric and topographic data into the model, from deep water to the inland limit of the study area of the Now Jade resort. The general mesh includes part of the coral reef and surrounding area, in order to avoid the possible effects of the nourishment to the ecosystem and adjacent beaches (Figure 3a). Secondly, a higher resolution mesh was constructed for a more accurate simulation of the processes directly induced by the project (Figure 3b). Squared, 8 m long cells were used for the numerical modelling in the general mesh; whilst a cell size varying from 15 to 1 m was selected for the computation of the fine grid mesh. The study on the general mesh was carried out through the comparison of the results for two different seabed configurations with and without the micro-Sand Engine.

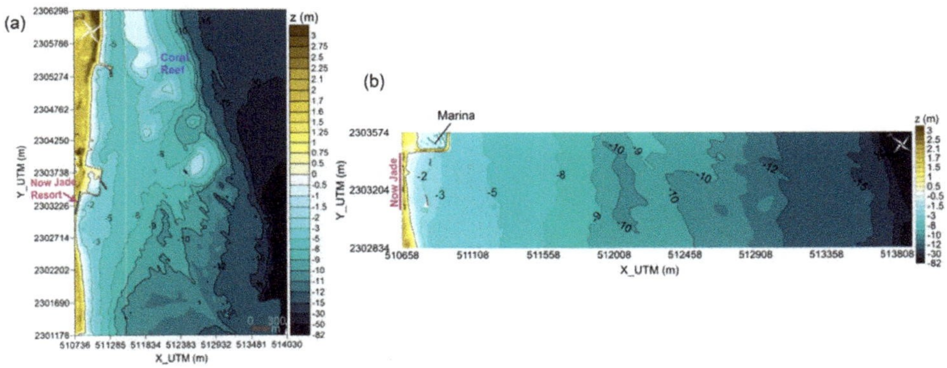

Figure 3. Bathymetric data within the domain of the simulation grid meshes for XBeach model: (**a**) general mesh; (**b**) detailed mesh.

4. Results

4.1. Beach Evolution & Diagnosis

The evolution of the dry beach and the construction works of the hotel is illustrated in satellite images from 2004 to 2017 (Figure 4). Of note are a small dune extending longitudinally along the coast before the hotel was built (see image from 2004 in Figure 4 and the dune in Figure 2d), and the reduced width of the dry beach in the northern half of the study area at various dates in the period.

The quantification of the total sediment balance given by topographic surveys (LEICA 3D scanner with 1–2 mm resolution) on 30 October 2012 (Figure 5a) and 26 April 2017 (Figure 5b) gave a total loss of 77 m^3, with a maximum value of 516 m^3 in the central area of the beach (highlighted by a rectangle in Figure 5c). Approximately 7 m in width and 0.9 m in elevation were lost in the most critical beach section (section P-P′ in Figure 5c,d), located 130 m from the northern end of the beach, reaching a minimum width of 9 m. The gain of sediment observed (in blue in Figure 5c), principally in the southern area, was related to the protection offered by two submerged breakwaters, constructed in 2010 and 2012 at 2 m depth, of 60 and 45 m long in the south and north, respectively, of the beach [37] (Figure 2c). Despite the protection of these structures, chronic erosion was still produced on the beach in subsequent years, especially in spring (April), as shown in the satellite images from 2015 to 2017 (Figure 4). This situation frequently leads to a beach state that is less desirable for recreational use by tourists in the high-season (personal communication with the hotel manager), and increases the flooding risk under the dominant wave climate of the region.

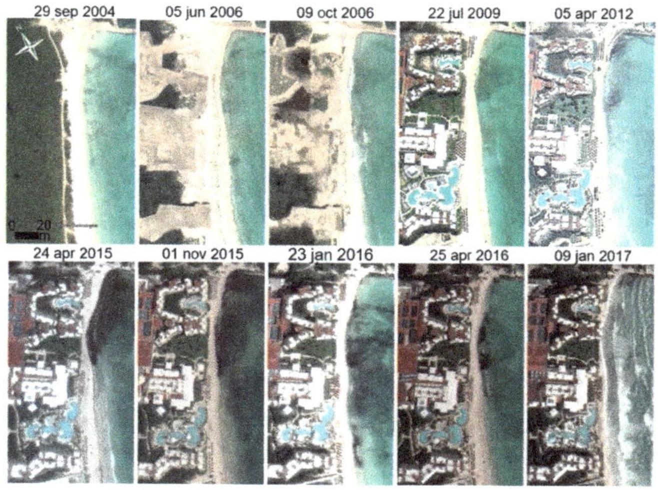

Figure 4. Evolution of the dry beach from 2004 to 2017 (Source: Google Earth).

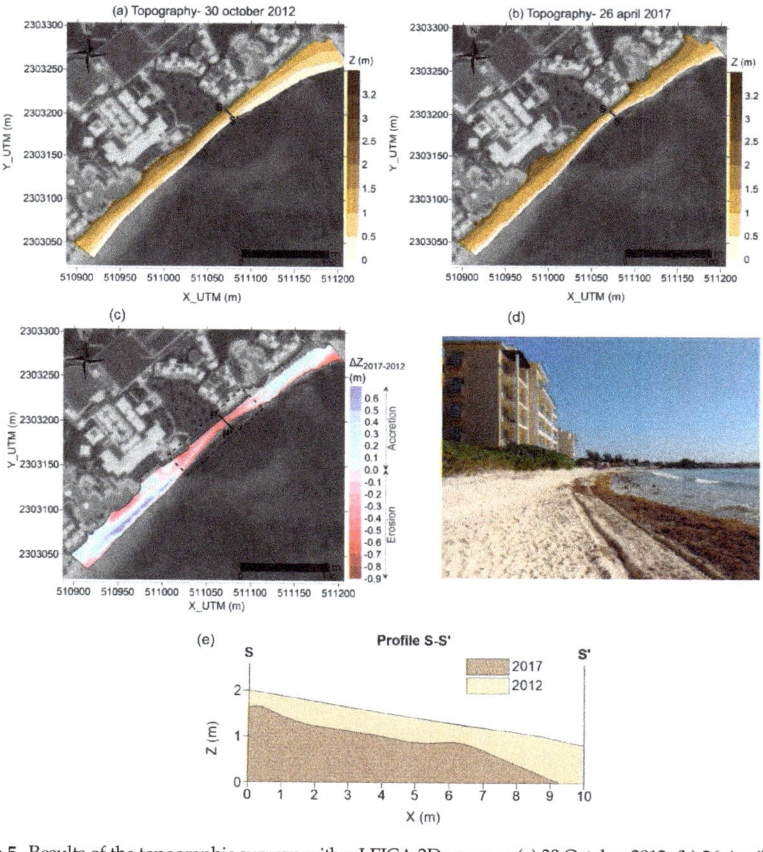

Figure 5. Results of the topographic surveys with a LEICA 3D scanner: (**a**) 30 October 2012; (**b**) 26 April 2017; (**c**) variation of the elevation within the period; (**d**) view of the beach section most eroded in 2017 (section P-P′ in Figure 5c); (**e**) Transverse profile S-S′ to show the sediment lost in the dry beach section most affected.

4.2. The Micro Sand Engine Project

Given the importance of taking measures to protect the coastal zone and preserving the environment in the study area, the innovative strategy called Micro-Sand Engine was proposed. The concept is based on massive Sand Engine nourishment, but on a very small spatial scale. The novel solution is intended to work in fragile areas. A small, artificial beach nourishment would be quick and easy to perform, and cost effective. It also avoids the intense sediment movement of traditional beach nourishment programs and thus prevents damage to the surrounding ecosystems (coral reef in this case). In turn, the natural distribution of the sand beach nourishment will be deposited in accordance with the natural shape and slope of the natural beach.

The project is complemented by building a small coastal dune, designed to provide a sand reservoir and thus strengthen the beach profile.

Design of the Micro Sand Engine and Sand Dune

Although the 'hook' was the preferred Sand Engine shape, the ease of construction and the small dimensions of the beach led to the 'bell-shape' being used in the study area. The salient dimensions are 33 m long in transversal direction (reaching 0.5 m depth), and 45 m alongshore, covering the beach section most affected by erosion (Figure 6). A maximum elevation of 0.8–1.1 m above the mean sea level was defined for the micro Sand Engine, considering the historical beach evolution (see 'Section S1' and 'Section S2' in Figure 7).

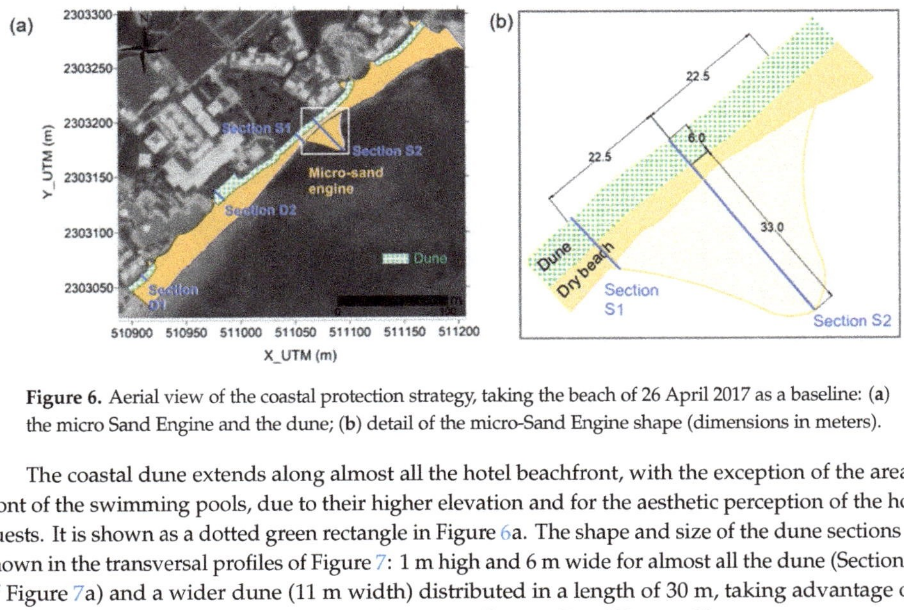

Figure 6. Aerial view of the coastal protection strategy, taking the beach of 26 April 2017 as a baseline: (**a**) the micro Sand Engine and the dune; (**b**) detail of the micro-Sand Engine shape (dimensions in meters).

The coastal dune extends along almost all the hotel beachfront, with the exception of the area in front of the swimming pools, due to their higher elevation and for the aesthetic perception of the hotel guests. It is shown as a dotted green rectangle in Figure 6a. The shape and size of the dune sections are shown in the transversal profiles of Figure 7: 1 m high and 6 m wide for almost all the dune (Section D1 of Figure 7a) and a wider dune (11 m width) distributed in a length of 30 m, taking advantage of a wider beach section to create a larger sand reservoir (Section D2 of Figure 7b).

The sediment source for the project, as mentioned previously, is very close to the breakwater alongside the marina (B, in red, Figure 2c). It is important to note that the shoal formed there is 150 m from the shore; far enough offshore for it not to travel naturally to the beach, but close enough to easily be pumped to the micro Sand Engine, from a depth of around 2–2.5 m. A sediment compatibility analysis was performed to validate the characteristics of the borrowed sediment as filling material. Seven sediment samples were taken at different locations along three transverse beach profiles (Sections S1, S2, and S3 in Figure 2c). The grain size of the borrowed sediment was slightly finer than the native sample at the micro Sand Engine site (Section S2 of Figure 2c), with a D50 of

0.220 mm vs. 0.267 mm, but appropriate considering a overfill factor of 1.3. A total nourishment volume of 2443 m^3 would be required for the project, of which 964 m^3 would be used for the micro Sand Engine and 1479 m^3 for the dune sections.

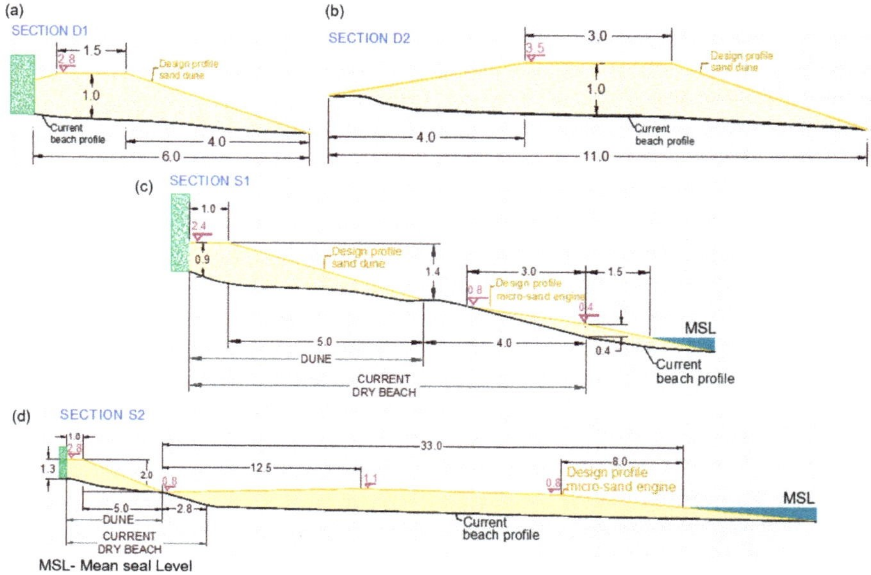

Figure 7. Beach profile designs: (**a**,**b**) coastal dune; (**c**,**d**) micro Sand Engine.

The monitoring program of the project includes a monthly survey of beach profiles to determine the necessity of a new beach nourishment.

The morphological evolution of the micro Sand Engine, which is expected to be constructed in 2020, was assessed numerically, for the mild and energetic wave conditions typical of the area. The results are presented in Section 4.3.2.

4.3. Modeling the Morphological Response of the Micro Sand Engine

4.3.1. Modeled Wave Scenarios

The wave roses presented in Figures 8 and 9 and the joint probability of Hs (significant wave height), Tp (peak period), and Dir (wave direction), in Figure 10, show calm conditions that dominate throughout the year, with a significant wave height of less than 1.6 m, coming from the east-southeast direction (Figures 8a and 10b), and peak periods of 6 to 8 s (Figure 10a). In storm conditions, waves from the east to the southeast sectors were frequent, with peak periods of 7 to 10 s (Figures 8b and 10), though also with a prevalence of those in the east-southeast. In the seasonal wave roses (Figure 9), storm waves from the north-northeast are also present in Autumn due to 'Nortes', though with very a low frequency of occurrence.

The seasonal wave roses show the storms typical of the area: (1) waves from the east-northeast, in autumn and winter, due to cold fronts; (2) frequent eastern storms throughout the year, except in the summer; (3) storms producing waves from the east-southeast (more frequent in spring and winter) and southeast (principally in spring) (Figure 9).

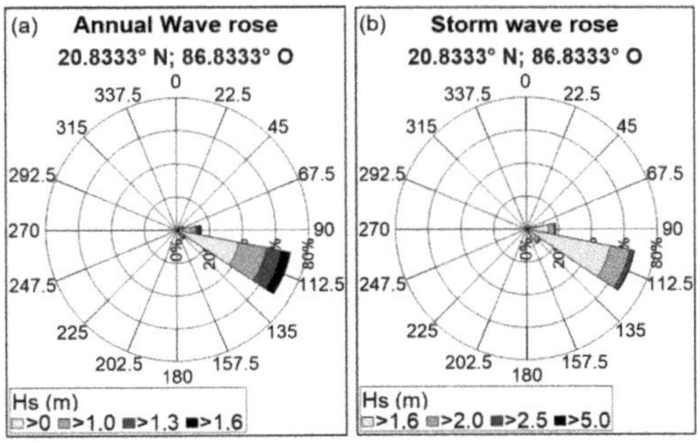

Figure 8. Wave roses: (**a**) annual; (**b**) storm conditions.

Two wave scenarios were chosen as representative of calm, or non-storm, wave conditions: Hs= 1 m and peak period, Tp, of 8 s coming from the east and the east-southeast. To analyze the effects of typical storms of the region, four simulation scenarios of the most energetic waves were modelled (see Table 1). These scenarios did not consider hurricanes, since after the passage of these extraordinary wave and wind phenomena, substantial reconstruction work on the Sand Engine and the dune would be required. Sea level variation was not taken into account for the simulation scenarios due to the negligible astronomical tide and storm surge values of typical storms.

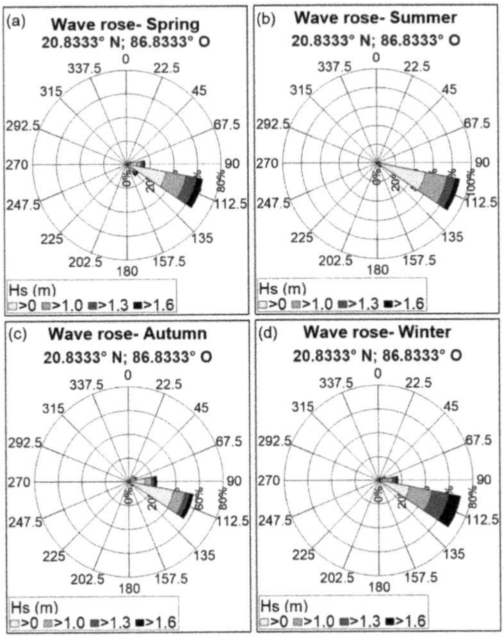

Figure 9. Seasonal wave roses: (**a**) Spring; (**b**) Summer; (**c**) Autumn; (**d**) Winter.

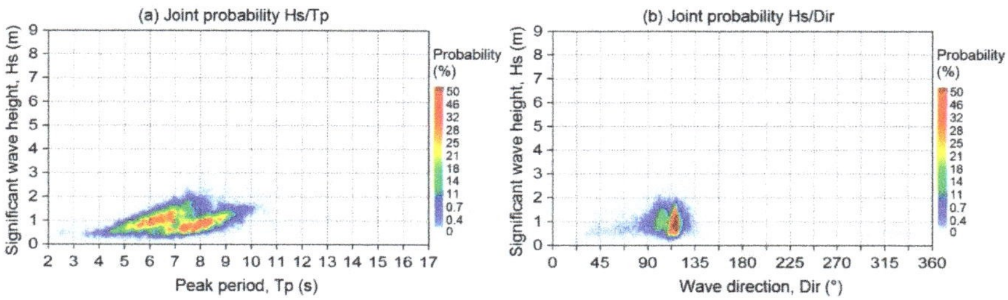

Figure 10. Joint probaility Hs, Tp, Dir: (**a**) Hs/Tp; (**b**) Hs/Dir.

Table 1. Definition of the numerical simulation wave scenarios.

Scenario	Significant Wave Height Hm0 (m)	Peak Period Tp (s)	Wave Direction
Calm or non-storm conditions	1	8	E ESE
Storm conditions	2	8 7	ENE E ESE SE

4.3.2. Model Results

Using the general area grid, the numerical modeling predicted changes in the beach morphology only limited to the area of the nourishment. No effects were seen for nearby beaches or the coral reef. This is as expected, given the small scale of the project and the northward constraint to sediment movement of the marina. The fine grid mesh calculations show relatively high waves (~1 m) that reach the area in storm and even calm wave conditions. In particular, waves coming from the east-southeast and southeast directions seem to be the worst cases (Figure 11).

The marine processes redistribute the sediment along the beach, according to the wave energy and wave-induced currents that impact the Sand Engine. An example of the current field around the Sand Engine in its initial condition for a 'Norte' storm can be seen in Figure 12a. The morphological change due to these conditions after 10 h has been drawn in Figure 12b, and in Figure 12c, after 30 h. The computed sediment transport will benefit the narrowest beach sections of the hotel beachfront.

Figure 12 clearly shows that as the Sand Engine erodes, the adjacent beach areas benefit from the sediment distribution. In turn, the whole beachfront is expected to rise in elevation, improving its stability. Once the sand from the Sand Engine has been totally distributed, strong sea states could damage the beach, hence the need to reconstruct it periodically (it is foreseen that reloading the Sand Engine will be necessary every year, or once every two years).

The degree of beach changes was more important with waves coming from the east-northeast, east, east-southeast, and southeast, respectively. It is interesting to note that the micro Sand Engine is eroded rapidly, but that the shape of the coastline becomes stable, keeping a small sand salient. Figure 13 shows that even for storm conditions (Figure 13b), the stable salient is present. Obviously, after the occurrence of several different sea states, particularly very intense ones, the salient will be removed and the beach will have to be reloaded. It is expected that maintenance of the Sand Engine will be needed annually, or biannually. Figure 14 shows the beach profile for Section S2 (Figure 7d). The final elevation of the profile is seen to be similar in calm and storm conditions, confirming that the salient remnant is stable and the beach is likely to remain sound for a determined time span.

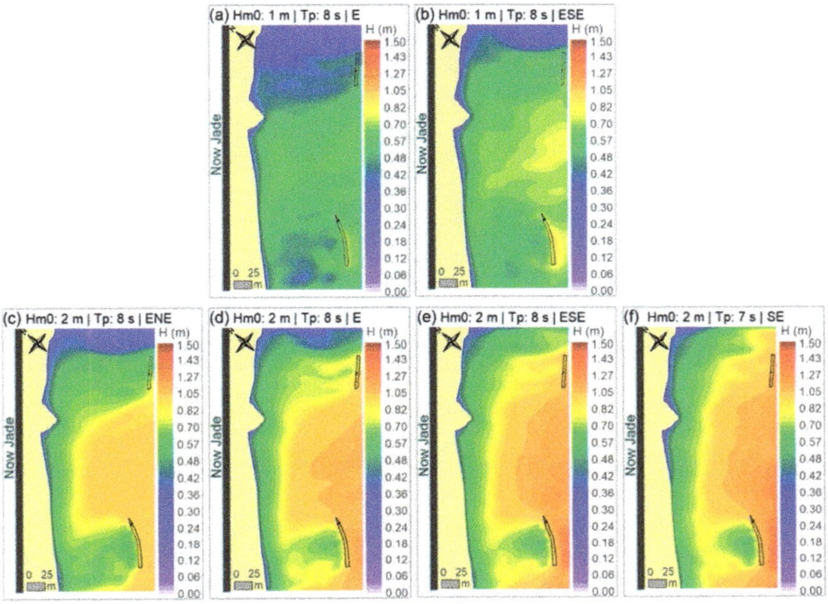

Figure 11. Spatial distribution of wave heights for the scenarios simulated: (**a**,**b**) non-storm wave conditions; (**c**–**f**) storm wave conditions.

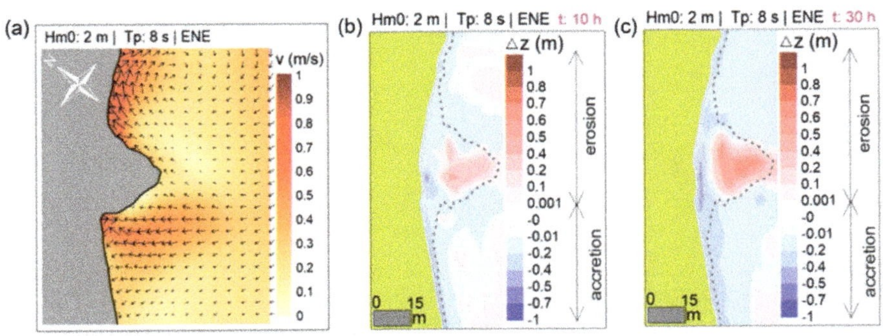

Figure 12. Current field and morphological change of the micro Sand Engine for a 'Norte' storm (blue: accretion; red: erosion): (**a**) circulation of wave currents; (**b**) morphological change after 10 h of simulation; (**c**) morphological change after 30 h of simulation.

The computed morphological development of the project shows how the beach slope becomes milder, which is preferable for tourist activities, as the sand spreads north and south, (Figure 12b,c). As the sand of the nourishment is deposited onshore, the dry beach increases in the sections adjacent to the micro-Sand Engine, by a similar magnitude for the six wave directions, once stability is reached (Figure 13).

The numerical model also showed the important role of the sand dunes for cross-shore sediment exchange from the beach to the dune, particularly for the ESE and SE storm scenarios (Figure 14e,f) (see the variation of the beach profiles in Figure 14).

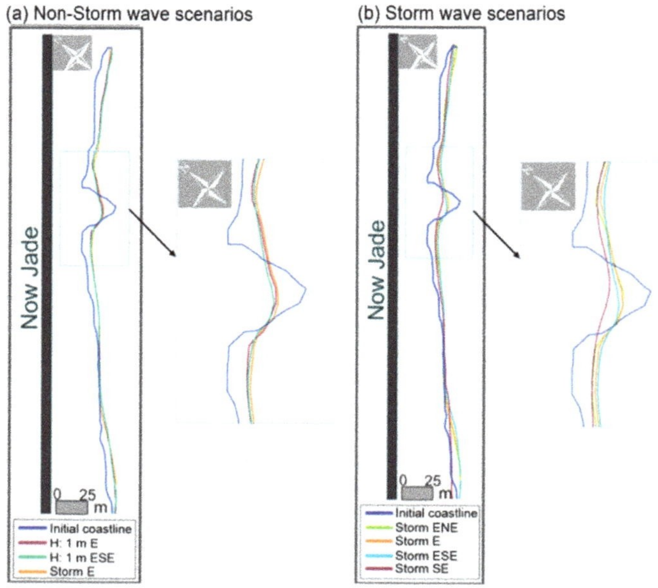

Figure 13. Coastline changes for the simulated scenarios (t: 30 h): (**a**) non-storm wave conditions; (**b**) storm scenarios.

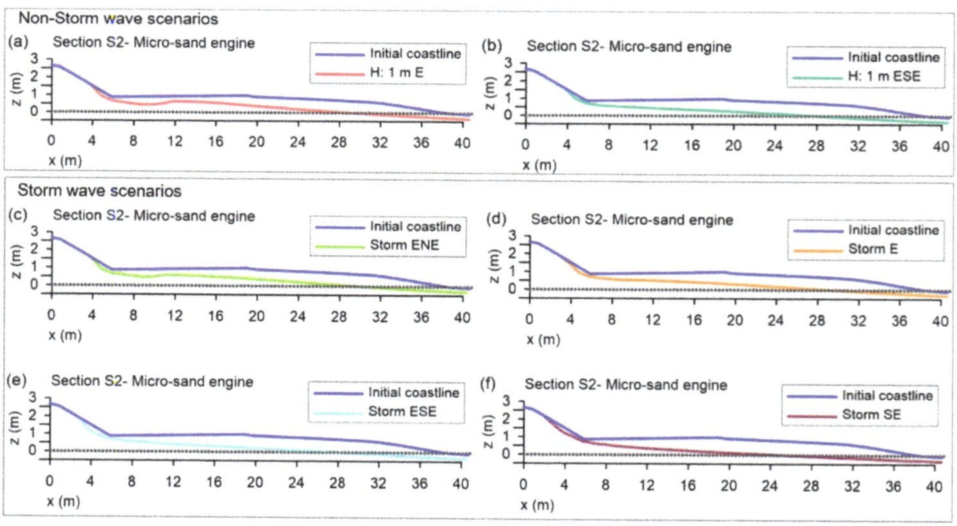

Figure 14. Variations in elevations of the transverse profiles, defined by the symmetry axis of the micro Sand Engine, (Section S2 in Figure 7d) (t: 30 h): (**a**,**b**) non-storm wave conditions; (**c**–**f**) storm scenarios.

5. Discussion and Conclusions

Beach nourishment has been recognized as the preferred coastal protection alternative for more than 30 years [38]. Undoubtedly, the successes of mega-nourishments encourage the implementation of other large-scale projects [39]. Huge beach fills also offer benefits such as long useful lives and reduced maintenance frequency. These engineering works consume time, space, and require the management of heavy machinery both in the sea and on land. These requirements are not always feasible in recreational beaches, where the work must be planned to interrupt as little as possible tourist activities, or even be carried out alongside these. In such cases, the work should entail the least invasive activities possible and use equipment which is as light as possible. Unfortunately, little research has yet focused on understanding tourism in the context of the coastal environment, [40] nor on the impact on tourism of coastal protection activities. The micro Sand Engine proposed here is compatible with the needs of the tourism industry, as it can be performed quickly and with relatively small machines.

The micro Sand Engine performance was computed using the XBeach model. The Xbeach has shown good results in estimating coastal morphodynamics related to beach fills [41,42]. The results obtained in this work show quite a dynamic bell-shaped feature that rapidly cedes sand to its surrounding area. It has to be noted that the Xbeach model is known to overestimate erosion [43]; this means that the results of the micro Sand Engine fall on the safe side. Several other methods and models exist for the prediction of beach nourishment evolution [6,44,45]. Many of these models have shown good results, but as they were calibrated with measured data, their use is restricted to specific sites. Whenever a micro Sand Engine is constructed, a monitoring program is needed; then the numerical modelling could also be refined.

The analysis carried out in this paper shows that the Sand Engine strategy, carried out on a very small scale in Mexico, is an effective, economic, and sustainable measure for beach protection. Some of the benefits of the massive Sand Engine project, such as the environmental services of recreational functions of the beach, or its functioning as a sand reservoir to provide long-term coastal protection are also assumed to occur through this project.

The changes induced by this type of nourishment, concentrated in space and time, are expected to also show benefits in the damaged ecosystems nearby. Given the small spatial scale, with depths of less than 1 m with a mild beach slope, many of the disadvantages of traditional shoreface replenishments are expected to be avoided with the implementation of this project.

The XBeach numerical modeling showed that, under typical wave climate conditions of the region, the bell-shape micro Sand Engine in Puerto Morelos will redistribute the sediment along the beach at different speeds, depending on the direction and magnitude of the incident waves, which vary in frequency according to the seasons. The numerical results and the satellite images show that the most critical beach state was in spring, related to the higher frequency of storms from the east-southeast and southeast.

To validate the functioning of this protection measure, which will be greatly affected by other ecosystem restoration work, monitoring will be necessary. However, it is clear that the response of the micro Sand Engine is very dynamic.

The findings of this paper are encouraging for increased use of eco-friendly practices for coastal protection. The bell-shaped Sand Engine can be applied anywhere, worldwide, where small nourishments are required and maintenance work is feasible. It is hoped that this innovative solution can be a useful beach conservation strategy in places where limited sediment is available and where tourist activities have to be disrupted as little as possible.

Author Contributions: Conceptualization, M.E. and E.M.; methodology, M.E. and E.M.; software, M.E. and E.M.; validation, E.M. and R.S.; formal analysis, M.E.; investigation, E.M. and R.S.; resources, R.S.; data curation, M.E. and R.S.; writing—original draft preparation, M.E.; writing—review and editing, E.M. and R.S.; visualization, M.E.; supervision, E.M. and R.S.; project administration, E.M. and R.S.; funding acquisition, R.S. All authors have read and agreed to the published version of the manuscript.

Funding: This research was partially funded by Promociones Marina Morelos S de RL de CV under contract 182017.

Acknowledgments: The authors wish to thank Now Jade Riviera Cancun Resort, for providing data and access to its facilities, especially to Eduardo Solla and Pablo Lucena.

Conflicts of Interest: The authors declare no conflict of interest. The funders had no role in the design of the study; in the collection, analyses, or interpretation of data; in the writing of the manuscript, or in the decision to publish the results.

References

1. McKenna, J.; MacLeod, M.; Power, J.; Cooper, A. *Rural Beach Management: A Good Practice Guide*; Donegal County Council, Lifford, Co.: Donegal, Ireland, 2000.
2. Buisson, P.; Rousset, A.; ANCORIM; Massey, J. *ANCORIM—Overview of Soft Coastal Protection Solutions*; BRGM/ONF: Bordeaux, France, 2012; 56p.
3. Willems, G.; Abecasis, M.; Catena, M.M. *Inland & Maritime Waterways & Ports. Proceedings of the Technical Sessions*; Elsevier: Amsterdam, The Netherlands, 2013.
4. Gianou, K. *Soft Shoreline Stabilization: Shoreline Master Program Planning and Implementation Guidance*; Shorelands and Environmental Assistance Program, Washington Department of Ecology: Olympia, WA, USA, 2014; Publication No. 14-06-009.
5. Nguyen, V.T.; Vu, M.T.; Nguyen, T.V. Evaluation of Beach Nourishment Performance in Ba Lang Beach Using Numerical Modelling. In Proceedings of the International Conference on Asian and Pacific Coasts, Singapore, 25–28 September 2019; pp. 609–616.
6. Dean, R.G.; Yoo, C.H. Beach-nourishment performance predictions. *J. Waterw. Port. C-ASCE* **1992**, *118*, 567–586. [CrossRef]
7. Kuang, C.; Pan, Y.; Zhang, Y.; Liu, S.; Yang, Y.; Zhang, J.; Dong, P. Performance evaluation of a beach nourishment project at West Beach in Beidaihe, China. *J. Coast. Res.* **2011**, *27*, 769–783.
8. Walton, T.L., Jr.; Cheng, J.; Wang, R.; Manausa, M. Modeling of three beach fill projects. *Ocean Eng.* **2005**, *32*, 557–569. [CrossRef]
9. Karambas, T.V.; Samaras, A.G. Soft shore protection methods: The use of advanced numerical models in the evaluation of beach nourishment. *Ocean Eng.* **2014**, *92*, 129–136. [CrossRef]
10. Fricke, B.; Weilbeer, H. Numerical Modeling of Shoreface Nourishments in the Schleswig-Holstein Wadden Sea. *Coast. Struct.* **2019**, *2019*, 773–780.
11. Huisman, B.J.; Walstra, D.J.R.; Radermacher, M.; de Schipper, M.A.; Ruessink, B.G. Observations and modelling of shoreface nourishment behaviour. *J. Mar. Sci. Eng.* **2019**, *7*, 59. [CrossRef]
12. Wang, P.; Roberts, T.M.; Elko, N.A.; Beck, T.M. Factors controlling the first year performance of eight adjacent beach nourishment projects, west-central Florida, USA. In Proceedings of the ICCE 2008: 31st International Conference on Coastal Engineering, Hamburg, Germany, 31 August–5 September 2008; pp. 2532–2544.
13. Roberts, T.M.; Wang, P. Four-year performance and associated controlling factors of several beach nourishment projects along three adjacent barrier islands, west-central Florida, USA. *Coast. Eng.* **2012**, *70*, 21–39. [CrossRef]
14. Rivero, O.; Astrup, S.K. *Project Building with Nature (EU-InterReg)*; Ministry of Environment and Food of Denmark: Copenhagen, Denmark, 2018.
15. Waterman, R.E. Integrated Coastal Policy Via Building with Nature. Ph.D. Thesis, TU Delft, Delft, The Netherlands, December 2010.
16. De Vriend, H.J.; van Koningsveld, M.; Aarninkhof, S.G.; de Vries, M.B.; Baptist, M.J. Sustainable hydraulic engineering through building with nature. *J. Hydro Environ. Res.* **2015**, *9*, 159–171. [CrossRef]
17. Mulder, J.P.; Tonnon, P.K. Sand Engine: Background and design of a mega-nourishment pilot in the Netherlands. In Proceedings of the ICCE 2010: 32nd Conference on Coastal Engineering, Shanghai, China, 30 June–5 July 2010; Volume 30, pp. 3805–3814.
18. Mulder, J.P.M.; Stive, M.J. Zandmotor (Sand Motor): Building with nature. In Proceedings of the 25th ICID European Regional Conference, Integrated Water Management for Multiple Land Use in Flat Coastal Areas, Groningen, The Netherlands, 16–20 May 2011. Paper III-21. ICID.
19. Stive, M.J.F.; De Schipper, M.A.; Luijendijk, A.P.; Ranasinghe, R.W.M.R.J.B.; Van Thiel De Vries, J.S.M.; Aarninkhof, S.; Marx, S. The Sand Engine: A solution for vulnerable deltas in the 21st century? In Proceedings of the Coastal Dynamics 2013: 7th International Conference on Coastal Dynamics, Arcachon, France, 24–28 June 2013.
20. Van Slobbe, E.; de Vriend, H.J.; Aarninkhof, S.; Lulofs, K.; de Vries, M.; Dircke, P. Building with Nature: In search of resilient storm surge protection strategies. *Nat. Hazards.* **2013**, *66*, 1461–1480. [CrossRef]
21. Luijendijk, A.P.; van Oudenhoven, A. *The Sand Motor: A Nature-Based Response to Climate Change. Findings and Reflections of the Interdisciplinary Research Program Naturecoast*; TU Delft Publishers: Delft, The Netherlands, 2019.
22. Kaji, A.O. Assessment of the Variables Influencing Sediment Transport at the Sand Motor. Master's Thesis, Delft University of Technology, Delft, The Netherlands, 2013.

23. De Schipper, M.A.; de Vries, S.; Ruessink, G.; de Zeeuw, R.C.; Rutten, J.; van Gelder-Maas, C.; Stive, M.J. Initial spreading of a mega feeder nourishment: Observations of the Sand Engine pilot project. *Coast. Eng.* **2016**, *111*, 23–38. [CrossRef]
24. Luijendijk, A.P.; Ranasinghe, R.; de Schipper, M.A.; Huisman, B.A.; Swinkels, C.M.; Walstra, D.J.; Stive, M.J. The initial morphological response of the Sand Engine: A process-based modelling study. *Coast. Eng.* **2017**, *119*, 1–14. [CrossRef]
25. Luijendijk, A.P.; Huisman, B.J.A.; De Schipper, M.A. Impact of a storm on the first-year evolution of the Sand Engine. In Proceedings of the Coastal Sediments, San Diego, CA, USA, 11–15 May 2015.
26. Kaji, A.O.; Luijendijk, A.P.; Van Thiel de Vries, J.S.M.; De Schipper, M.A.; Stive, M.J.F. Effect of different forcing processes on the longshore sediment transport at the Sand Motor, the Netherlands. In Proceedings of the ICCE 2014: 34th International Conference on Coastal Engineering, Seoul, Korea, 15–20 June 2014.
27. De Schipper, M.A.; De Vries, S.; De Zeeuw, R.C.; Rutten, J.; Ruessink, B.G.; Aarninkhof, S.G.J.; Gelder-Maas, V. Morphological development of a mega-nourishment; first observations at the Sand Engine. In Proceedings of the ICCE 2014: 34th International Conference on Coastal Engineering, Seoul, Korea, 15–20 June 2014; Volume 34, pp. 1–6.
28. Kaergaard, K.; Droenen, N. A Hybrid Shoreline Model for the Sand Engine: Comparison with Observations and Long Term Predictions. In Proceedings of the The Proceedings of the Coastal Sediments, San Diego, CA, USA, 11–15 May 2015.
29. Wengrove, M.E.; Henriquez, M.; De Schipper, M.A.; Holman, R.; Stive, M.J.F. Monitoring morphology of the Sand Engine leeside using Argus' cBathy. In Proceedings of the Coastal Dynamics 2013: 7th International Conference on Coastal Dynamics, Arcachon, France, 24–28 June 2013.
30. De Vries, S.; Radermacher, M.; De Schipper, M.A.; Stive, M.J.F. Tidal dynamics in the Sand Motor lagoon. In Proceedings of the E-proceedings of the 36th IAHR World Congress, Hague, The Netherlands, 28 June–3 July 2015.
31. Hoonhout, B.; de Vries, S. Field measurements on spatial variations in aeolian sediment availability at the Sand Motor mega nourishment. *Aeolian Res.* **2017**, *24*, 93–104. [CrossRef]
32. Hagenaars, G.; de Vries, S.; Luijendijk, A.P.; de Boer, W.P.; Reniers, A.J. On the accuracy of automated shoreline detection derived from satellite imagery: A case study of the Sand Motor mega-scale nourishment. *Coast. Eng.* **2018**, *133*, 113–125. [CrossRef]
33. Coronado, C.; Candela, J.; Iglesias-Prieto, R.; Sheinbaum, J.; López, M.; Ocampo-Torres, F.J. On the circulation in the Puerto Morelos fringing reef lagoon. *Coral Reefs* **2007**, *26*, 149–163. [CrossRef]
34. Mariño-Tapia, I.; Enriquez, C.; Silva-Casarín, R.; Mendoza-Baldwin, E.; Mancera, E.E.; Ruiz-Rentaría, F. Comparative morphodynamics between exposed and reef protected beaches under hurricane conditions. In Proceedings of the ICCE 2014: 34th International Conference on Coastal Engineering, Seoul, Korea, 15–20 June 2014.
35. Silva, R.; Ruíz, G.; Posada, G.; Pérez, D.; Rivillas, G.; Espinal, J.; Mendoza, E. *Atlas de Clima Marítimo de la Vertiente Atlántica Mexicana*; Universidad Nacional Autónoma de México: Mexico City, Mexico, 2008. (In Spanish)
36. Roelvink, D.; van Dongeren, A.; McCall, R.; Hoonhout, B.; van Rooijen, A.; van Geer, P. *XBeach Manual. Model Description and Reference Guide to Functionalities*; UNESCO-IHE Institute of Water Education and Delft University of Technology: Delft, The Netherlands, 2015.
37. Silva, R.; Mendoza, E.; Mariño-Tapia, I.; Martínez, M.L.; Escalante, E. An artificial reef improves coastal protection and provides a base for coral recovery. *J. Coast. Res.* **2016**, *75*, 467–471. [CrossRef]
38. Dean, R.G. Additional Sediment Input to the Nearshore Region. *Shore Beach* **1997**, *55*, 76–81.
39. Brown, J.M.; Phelps, J.J.C.; Barkwith, A.; Hurst, M.D.; Ellis, M.A.; Plater, A.J. The effectiveness of beach mega-nourishment, assessed over three management epochs. *J. Environ.* **2016**, *184*, 400–408. [CrossRef]
40. Klein, Y.L.; Osleeb, J. Determinants of coastal tourism: A case study of Florida beach counties. *J. Coast. Res.* **2010**, *26*, 1149–1156. [CrossRef]
41. Williams, J.J.; Esteves, L.S.; Rochford, L.A. Modelling storm responses on a high-energy coastline with XBeach. *Model. Earth Syst. Environ.* **2015**, *1*, 14. [CrossRef]
42. Ruffini, G.; Briganti, R.; Alsina, J.M.; Brocchini, M.; Dodd, N.; McCall, R. Numerical Modeling of Flow and Bed Evolution of Bichromatic Wave Groups on an Intermediate Beach Using Nonhydrostatic XBeach. *J. Waterw Port C-ASCE* **2020**, *146*, 04019034. [CrossRef]

43. Elsayed, S.M.; Oumeraci, H. Effect of beach slope and grain-stabilization on coastal sediment transport: An attempt to overcome the erosion overestimation by XBeach. *Coast. Eng.* **2017**, *121*, 179–196. [CrossRef]
44. Work, P.A.; Dean, R.G. Assessment and prediction of beach-nourishment evolution. *J. Waterw Port. C-ASCE* **1995**, *121*, 182–189. [CrossRef]
45. Benedet, L.; Finkl, C.W.; Campbell, T.; Klein, A. Predicting the effect of beach nourishment and cross-shore sediment variation on beach morphodynamic assessment. *Coast. Eng.* **2004**, *51*, 839–861. [CrossRef]

Journal of
Marine Science and Engineering

MDPI

Article

Communicating Simulation Outputs of Mesoscale Coastal Evolution to Specialist and Non-Specialist Audiences

Andres Payo [1,*], Jon R. French [2], James Sutherland [3], Michael A. Ellis [1] and Michael Walkden [4]

1 British Geological Survey, Nicker Hill, Keyworth, Nottingham NG12 5GG, UK; mich3@bgs.ac.uk
2 UCL, Coastal & Estuarine Unit, Gower Street, London WC1E 6BT, UK; j.french@ucl.ac.uk
3 HR Wallingford, Howbery Park, Wallingford, Oxfordshire OX10 8BA, UK; j.sutherland@hrwallingford.com
4 WSP, Keble House, Southernhay Gardens, Exeter EX1 1NT, UK; mike.walkden@wsp.com
* Correspondence: agarcia@bgs.ac.uk; Tel.: +44-(0)115-936-3103

Received: 29 February 2020; Accepted: 20 March 2020; Published: 1 April 2020

check for updates

Abstract: Coastal geomorphologists and engineers worldwide are increasingly facing the non-trivial challenge of visualising and communicating mesoscale modelling assumptions, uncertainties and outcomes to both coastal specialists and decision-makers. Visualisation of simulation outcomes is a non-trivial problem because the more abstract scientific visualisation techniques favoured by specialists for data exploration and hypothesis-testing are not always as successful at engaging decision-makers and planners. In this paper, we show how the risk of simulation model outcomes becoming disconnected from more realistic visualisations of model outcomes can be minimised by using the Coastal Modelling Environment (CoastalME). CoastalME is a modelling framework for coastal mesoscale morphological modelling that can achieve close linkages between the scientific model abstractions, in the form of lines, areas and volumes, and the 3D representation of topographic and bathymetric surfaces and shallow sub-surface sediment composition. We propose and illustrate through the study case of Happisburgh (eastern England, UK), a transparent methodology to merge the required variety of data types and formats into a 3D-thickness model that is used to initialise a simulation. We conclude by highlighting some of the barriers to the adoption of the methodology proposed.

Keywords: visualisation; erosion; modelling; stakeholders

---

1. Introduction

One of the grand challenges facing coastal geomorphology today is to improve our ability to make quantitative predictions of morphological change at a scale that is relevant to longer-term strategic coastal management [1]. Following [2], this scale is herein referred to as the mesoscale, and is characterised by time horizons of the order 10^1 to 10^2 years and less rigorously imposed spatial dimensions of the order 10^1 to 10^2 km. Coastal engineers additionally face the challenge of achieving project approval which, amongst other activities, involves delivering such predictions of coastal change within an uncertainty framework that is robust enough to be useful to management and policy thinking [3]. Approval for coastal engineering schemes is granted only after the 'stakeholders'—a diverse group of people that typically includes representatives of governments, owners, directly and indirectly affected individuals and groups, and other individuals or groups who believe they may be impacted—have had the opportunity to express their reactions [3]. In this context, both coastal geomorphologists and engineers (i.e., specialists) increasingly face the non-trivial challenge of visualising and communicating the mesoscale modelling assumptions, uncertainties and outcomes to generally non-specialist decision-makers. A key aspect of this challenge is the overarching requirement that the application of any environmental model to decision making should be transparent and reproducible, as well as adequately representing the real system properties and behaviours [4–7]. Fulfilling these requirements often requires iterative communication at different stages of the simulation process.

Within the wider context of a simulation study [8], Figure 1 illustrates an idealised workflow for the simulation of mesoscale coastal geomorphological change to inform strategic coastal management and facilitate project approval. This includes seven main activities, four of which are development-related and three of which are concerned with the validation of the model. The four development activities are: (D1) knowledge acquisition; (D2) quantitative modelling (which integrates conceptual modelling, prototyping and model coding); (D3) experimentation, and; (D4) implementation. The outcome of each process is, respectively, a system description (i.e., Coastal System Maps (CSM) and Causal Loop Diagrams (CLD) [9,10]), a digital model, solutions to the problem modelled and/or a better understanding of the real world and improvements of the real world. The three validation activities encompass (V1) structural, (V2) behaviour and (V3) policy validity testing [11]. The outcome of each validation activity is an increase/decrease in the confidence level regarding the model scope and level of detail, model behaviour and the implications of the modelling results for policy, respectively. The double arrows in Figure 1 reflect the iterative nature of the process and the circular flow of these four main activities illustrates the potential to repeat the process of improvement through multiple simulations. In this work, our interest is on model content, in particular ensuring that the model has a sufficient level of detail for the behaviour validity testing (V2) and the policy validity testing (V3). It is useful to distinguish between the realistic and useful detail that can be included in relation to topography and sedimentology and the quite abstract representations of the processes governing geomorphological development that often underpin such mesoscale simulations.

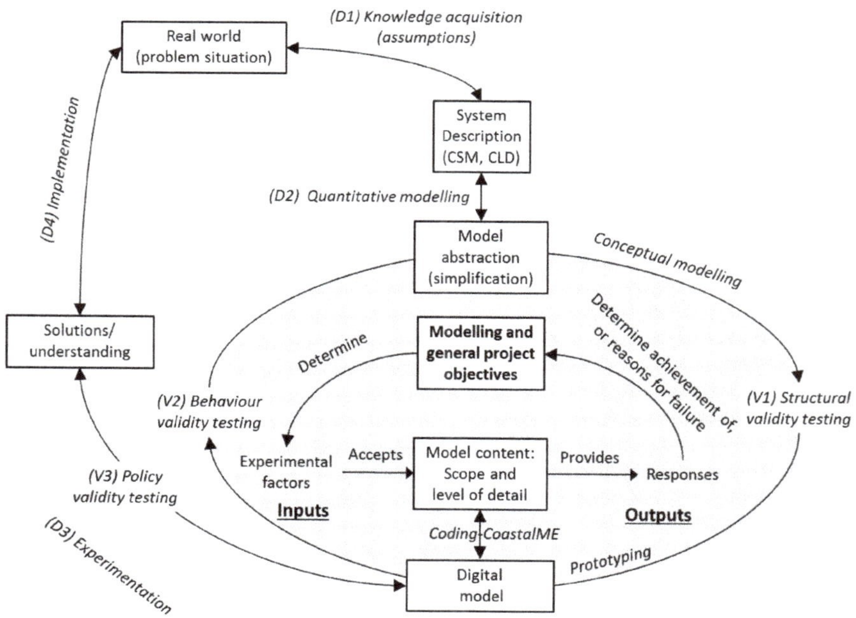

Figure 1. Idealised workflow for simulating mesoscale coastal geomorphological change to inform strategic coastal management, and to facilitate project approval (inspired by [8] and created by the authors).

Visualisation of simulation outcomes is a non-trivial problem because the more abstract techniques generally favoured by technical specialists for data exploration and hypothesis-testing are not always successful at engaging decision-makers and planners [12]. A general perception amongst planners and managers appears to be that the more detail within a visualisation, the more accurate and believable it is [13]. However, the inclusion of detail in visualisations of the outputs of relatively abstract models is problematic and potentially misleading. This raises the danger that the mismatch between the 'supply' of credible model data and the user-driven 'demand' for realistic information may lead to a divergence between the processes of model-based simulation and the visualisation of the outputs. This difficulty arises because the digital visualisation needs to be both scientifically credible but also provide sufficient detailed information at a level of realism that decision-makers believe is most suitable. Thus, although visualisation has become a valuable part of the geo-information toolkit, caution and agreed visualisation standards are still needed. Specifically, how do we best combine different knowledge and data sets into the modelling, and how do we best visualise the model outcomes for both the coastal specialist and non-specialist?

While sub-surface geology has long been recognised as an important factor for mesoscale coastal evolution [14], the problem of ensuring that the model has a sufficient representation of the sub-surface has often been approached in an ad-hoc way that makes it difficult to trace-back to the original geological data. For example, the modelling tool used by [12] to simulate the evolution of the soft cliff coastline of Norfolk, eastern England (UK), was the Soft-Cliff and Platform Erosion (SCAPE) model code [15], which uses data on geological composition and strength, among other input data. The relatively complex geology of the East Anglian coastline was characterised in the SCAPE model through the optimisation of rock strength and the scaling factor of a longshore sediment transport calibration parameter [15]. Informed by the cliff sediment size composition (i.e., proportions of mud, sand and gravel) provided by [16] and digitised historic maps, the SCAPE model was calibrated to accurately represent first an 87-year hindcast of cliff toe evolution [15] and then a longer 117-year simulation [17]. This model was later used to predict cliff change over 100 years under a range of potential coastal management strategies for North Norfolk Council, as part of the 2013 Cromer to Winterton Ness Coastal Management study [18] to inform the selection of a preferred coastal management strategy. In that example, the mismatch between the abstract representation offered by the SCAPE model and the end-user desire for visualisations with plausible detail was met by superimposing the projected recession on the observed baseline cliff position, retaining some of the detail of that cliff-line. In this way, a mismatch was introduced between the model output and the basis of the visualisation.

Cliff sediment yields [16] were estimated from the lithology exposed at the cliff face in 1995 and represent the yields of a 1m recession of the cliff face. This was assumed, without detailed empirical evidence, to be a good representation of the inland geology. As shown by [19], it is now possible to more accurately estimate the sediment yields of an eroding cliff and shore platform. Reducing the uncertainty on cliff yields is key as they determine the beach volume, which is related to the rate of cliff recession (i.e., annual cliff top recession rate decreases exponentially with the beach volume) [20]. In addition, the inclusion of more complete data on known geology and its sediments provides a means of introducing realistic localised detail into model simulations and the visualisation of the resulting projections, which can help to satisfy the demands of end-users.

The aim of this paper is to illustrate how closely linking the outputs from the Coastal Modelling Environment (CoastalME) [16] and Geographical Information System (GIS) data structures and output formats minimises the mismatch between scientifically sound model results and user-demanded realism in terms of how the coast is represented. To demonstrate the communication of complex beach-shore platform-cliff interactions to both specialist and non-specialist audiences, we have used the simulation outcomes obtained by [19] for the eroding coast of Happisburgh (eastern England, UK). Happisburgh provides an excellent example of how superficial geology mediates the observed rapid acceleration in cliff retreat after the removal of coastal defences [21,22]. We then briefly present CoastalME as an appropriate framework to create realistic representations of coastal landform complexes including sub-surface and digital elevation data. In particular, we present the CoastalME data structure and the workflow for the initialisation of a coastal simulation model. Then we briefly describe the workflow and simulation outcomes followed by [19] when simulating one year of the evolution of Happisburgh. We then illustrate different ways of visualising the simulation outcomes, from traditional cliff lines and profiles to evolving DEMs. Finally, we discuss how this novel workflow and visualisation is an improvement relative to previous and current practices.

2. Materials and Methods

2.1. Happisburgh Case Study Description

Happisburgh is located on the soft sediment coast of Norfolk, eastern England (Figure 2). The length of the coast that is simulated is ca. 3 km. The site is exposed to southern North Sea waves, with average annual significant wave heights (*Hs*) of 0.9 m and peak periods (*Tp*) of 4 s from the N-NNE. The wave climate is non-seasonal with similar moderate-energy summers (July to September, *Hs* = 0.95 m and *Tp* = 4 s) and moderate-energy winters (October to June, *Hs* = 0.92 m and *Tp* = 4 s), and extreme wave heights exceeding *Hs* = 6 m and *Tp* = 10 s. The coast is macro-tidal, with a mean spring range of 4.2 m and mean neap range of 2.1 m. Relative sea levels at this location have been rising for millennia, and under natural conditions, this part of the coast is erosional.

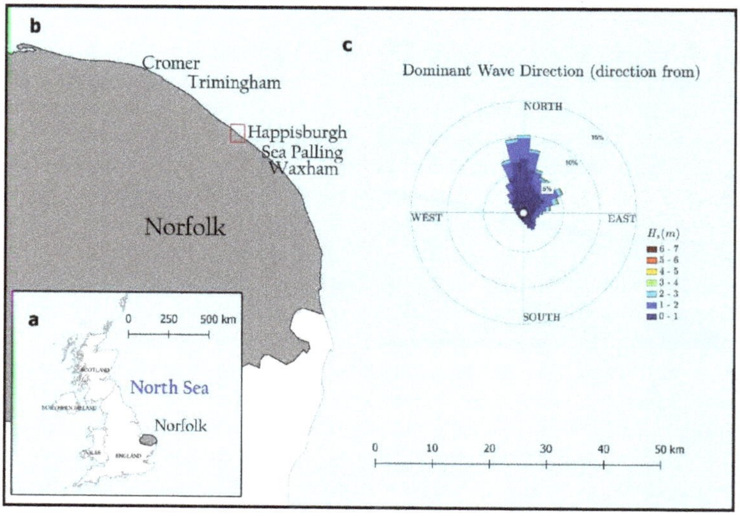

Figure 2. Study location: (**a**) Norfolk (grey polygon) on the east coast of England; (**b**) study site location (red rectangle) and nearby locations mentioned in this paper; (**c**) wave rose for Happisburgh created using downscaled data (1961–2016) from the Climate Projections 2009. Available at: http://ukclimateprojections.defra.gov.uk/ (accessed 30 Nov 2019).

Defences were constructed in 1958/59 (wooden revetments and steel sheet piles) and 1968 (wooden groins) [23,24] (Figure 3) on the soft cliff coastline, which formerly eroded at < 1 m/yr [24]. The steel sheet piles are at the landward limit of the groins that lie perpendicular to the coast with a length of 100 m and are spaced along the coast at intervals of 170 m. After construction of the defences, the rate of erosion decreased, and any subsequent loss of land was caused by the failure of unstable cliff slopes. From the late 1980s, the defences were not maintained, in part because of a lack of agreement regarding coastal protection, and in part, because of a lack of funding [23]. By 1991, defence failure led to selective defence removal on safety grounds along 900 m of coast [24], while adjacent defences remained. Subsequently, where defences were removed, excessive retreat occurred and over a period of 14 years, the cliff eroded on average 100 m landward, creating a shallow embayment. Between September 2001 and September 2003, 3.6×10^4 t of sediment was eroded from a 100 m section of the cliff [25] with retreat rates recorded between 8 and 10 m/yr. In 2007, with financial support from the local authority, the local community helped fund rock armouring along the cliff base to slow the erosion [26].

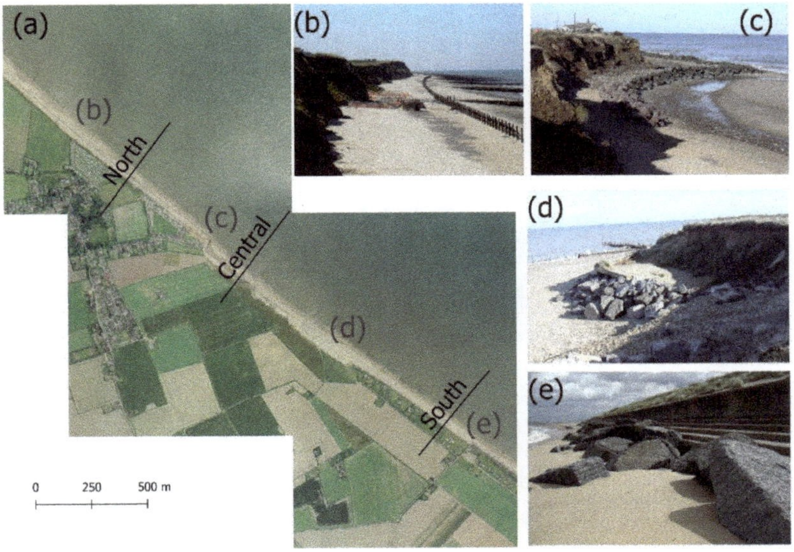

Figure 3. Human interventions along the Happisburgh coast; (**a**) aerial image from 2006 showing the transect locations (source: Aerial photography© UKP/Getmapping Licence No. UKP2006/01); (**b**) wooden revetments, steel sheet pile and groins in the northern part; (**c**) & (**d**) rock armouring at north-end and south-end of the embayment; (**e**) rock armouring in front of the seawall at the southern part.

At the southern end of the embayment, a sea-wall protects the low-lying farmland and tourist area of Eccles-on-Sea (Figure 3). This sea-wall was constructed after catastrophic floods in 1953 and in stages up to 1987. There are now three main elements making up the sea defence system: the beach, the sea wall and the sand dunes [27]. The beach becomes highly mobile during storms and can be drawn down to such an extent that the sea wall becomes unstable. In the early and mid-1990s, beaches in the Eccles/Sea Palling area reached critical levels where the sea wall foundations started to fail. Three emergency works contracts were implemented for placing rock protection along the toe of the sea wall. If the sea wall was allowed to collapse, the sand dunes would offer the last line of defence and would be breached rapidly by wave action [27]. Since 1996, the Environment Agency (EA) has undertaken a series of beach nourishments (around 150,000 m^3/yr on average [27]) at Sea Palling, about 5 km to the south and down-drift of Happisburgh. This nourishment aims to offset the concomitant reduction in sediment supply from cliff erosion along the Happisburgh-Trimingham coastal section and to maintain sea defence.

The cliffs at the study site have slopes slightly lower than and within the range of their peak angle of friction (i.e., 24 to 32 degrees [28]) and range in height from 6 m to 20 m above Ordnance Datum (OD) (Figure 4). They are composed of a sequence dominated by glacigenic sediments including multiple diamictons (admixtures of poorly-sorted clay, sand and gravel), separated by beds of stratified silt, clay and sand [29–31]. These units were deposited during several major incursions of glacier ice into the region during the Middle Pleistocene [32]. The basal unit that crops-out discontinuously at the base of the cliffs is the Crag Group (Early to early Middle Pleistocene age), which is typically obscured by modern beach material but is periodically exposed following storms [25]. The Crag Group consists of stratified sands and clays interpreted as shallow marine to inter-tidal in origin and punctuated by occasional elongated lenticular bodies of fluvial muds [33,34]. The Crag Group rests unconformably on an undulating upper surface of Chalk bedrock (Cretaceous age). Chalk bedrock depths increase from about −25 mOD at the northern end to −40 mOD at the southern end. The large potential stock of sand and gravel at the shore platform present in the Crag Group suggest that the beaches at Happisburgh might not be sediment starved, providing that the waves and currents have enough energy to erode and transport this material towards the coast.

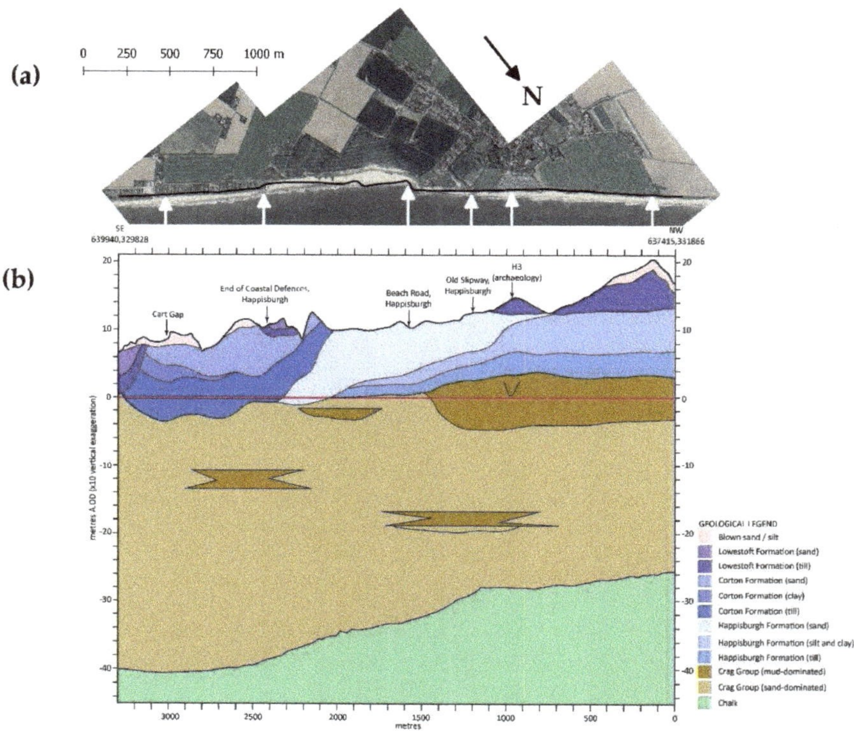

Figure 4. Main geological units at the study site along the 1999 cliff top line: (**a**) cliff top line of the year 1999 is shown as a solid black line on top of the year 2010 aerial photography of the study site; (**b**) main lithological units. Key landmarks along the cliff cross-section are named in (**b**), approximate locations on (**a**) are indicated by white arrows. Across-shore distances are distances measured along the cliff top line (starting at the northern end) and vertical elevation are relative to Ordnance Datum (which is approximately at mean sea level for the study site). For clarity, the vertical scale has been exaggerated 10 times.

2.2. CoastalME: Concept and Data Structure

CoastalME [35], is a framework to integrate coupled mesoscale reduced complexity models, reductionist coastal area models, data-driven approaches, and qualitative conceptual models. Integration of these heterogeneous approaches gives rise to model compositions that can potentially resolve decadal- to centennial-scale behaviour of diverse coupled open coast, estuary and inner shelf settings. However, coupling existing software models is not a trivial task. One approach involves the coupling of landform specific simulation models (e.g., cliffs, beaches, dunes and estuaries) that have been independently developed [36]. An alternative approach is to capture the essential characteristics of the landform-specific models using a common spatial representation within an appropriate software framework [35]. The latter avoids the problems that result from the model-coupling approach arising from between-model differences in the conceptualisations of geometries, volumes and locations of sediment. CoastalME should be understood as a software framework containing a minimum set of objects and a data structure suitable to integrate any coastal evolution simulation to create a digital model composition (i.e., CoastalME is a tool to create model compositions).

In CoastalME, the shoreface is conceptualised (Figure 5) as a set of sediment sharing cells interconnected by the alongshore sediment transport. Coastal morphological change is simulated as dynamically linked line and raster objects. The hierarchy of panels in Figure 5 illustrates how a real coastal morphology (upper panel) is conceptualised in terms of shoreline, shoreface profiles and estuary elements (middle panel). All elements can share sediment among them (double-headed arrow). The shoreface comprises both consolidated and non-consolidated material that forms the cliff, shore platform and beach, respectively (bottom panel). At every time step, the shoreline is delineated at the intersection of the sea level and the ground elevation. Shore face profiles are defined perpendicular to the shoreline. The sea level and wave energy constrain the proportion of shoreface profiles that are morphologically active at each time step. Eroded sediment from the consolidated profile is added to the drift material to advance the shoreline or loss as suspended sediment. Gradients of the littoral drift further control the advance and retreat of the beach profile and the amount of sediment shared with nearby sections of the coast.

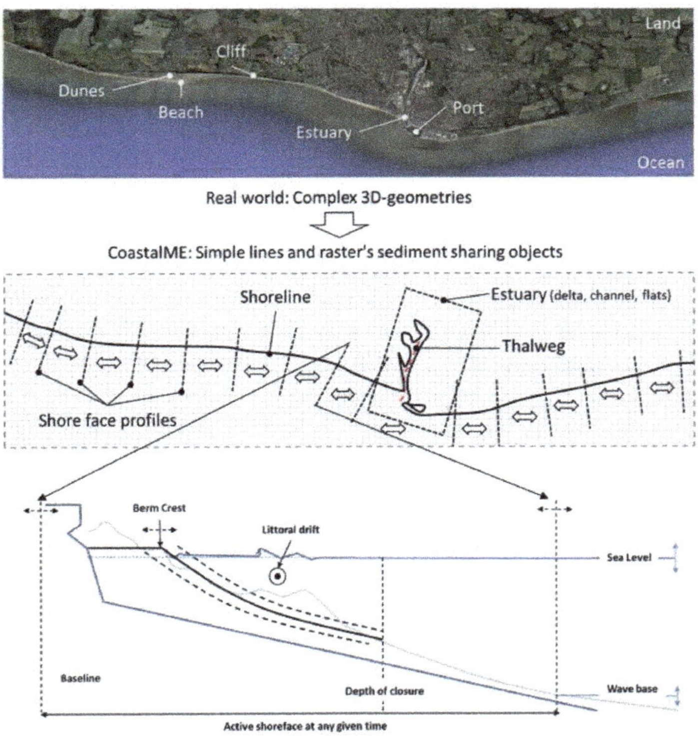

Figure 5. CoastalME modelling approach concept (source; [35]).

Figure 6 illustrates CoastalME topography and subsurface data structure. Ground elevation is characterised as a set of regular square blocks. The size of the square blocks is determined by the user based on the resolution of data availability and model outputs sensitivity. Each block has a global coordinate x, y, z. Each block might be composed of six different sediment fractions made of coarse sand and fine sediment sizes. Each sediment size fraction can be in a consolidated or unconsolidated state.

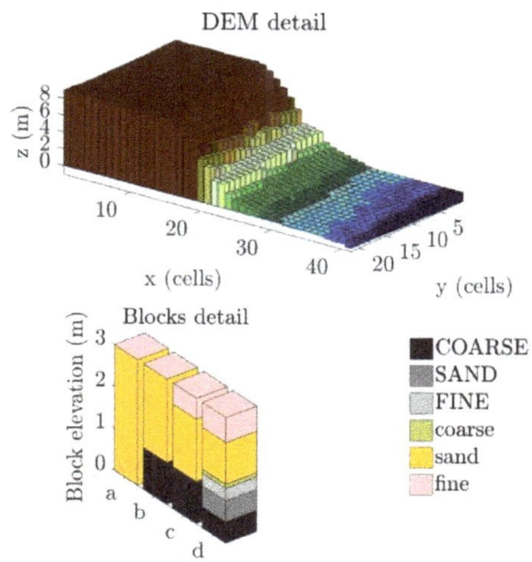

Figure 6. CoastalME data structure. Block types a, b, c and d illustrate blocks of same total elevation but with different sediment composition. Consolidated and unconsolidated sediment are represented by brighter-colours/non-capitalized-text and greys-colours/capitalized-text, respectively. (source; [35])

Input parameters for CoastalME are supplied via a set of raster, vector, and time series files and a text-format configuration and steering file [37]. Raster files represent the initial ground elevation, sediment thickness and coastal intervention. Vector files specify the locations within the model spatial domain for which wave (wave height, direction and period) forcing time series are provided. Tidal water surface elevation is also provided as a time series input file. CoastalME output consists of GIS layer snapshots, a result summary file, and a number of time-series files. The GIS files include both raster layers such as digital elevation models (DEMs) and sediment thickness and vector layers such as the coastline.

2.3. Simulation Outcomes of Happisburgh Annual Evolution

To illustrate the different visualisation options, we have used CoastalME simulation outcomes [19] of the relative contribution of back wearing and down wearing cliff and platform erosion to the near-shore sediment budget under idealised annual forcing conditions at Happisburgh (Table A1). This is based on a 360-day simulation at a 1-h numerical time step, with results saved after 1, 30, 60, 90, 120, 150, 180, 210, 230, 260, 290, 310, 340, 360 days. In [19], only the final beach thickness and elevation were shown. For this work, we have used the outcomes obtained for the simulations in best agreement with observations, which were obtained when the rock strength and hydrodynamic constants for the platform and cliff were set equal to 8×10^4 [$m^{9/4}s^{2/3}$] and 8×10^2 [$m^{9/4}s^{2/3}$], respectively. The CoastalME-Happisburgh simulation produced 73 different types of outcomes (Table A2), which includes 11 Comma Separated Value (CSV) files, 14 GIS vector files and 48 GIS raster files. The raster and vector files are snap-shots at user-defined intervals, while CSV files contain time series for all time-steps (here 86,400 representing 360 days at a 1-h interval).

Figure 7 illustrates the overall workflow followed by [19] to obtain the thickness model used for the idealised Happisburgh simulation. Payo et al. [16] first generated a 3D thickness model using three data sets (topography, bathymetry and subsurface lithology, using existing databases) as required by the CoastalME data structure, although, the proposed workflow is transferable to any other place where equivalent data sets are available. Firstly, the topography and bathymetry database are combined to create a seamless topo-bathymetric Digital Terrain Model (DTM) of the study site. For Happisburgh, this combined Environment Agency LIDAR DTM data for 1999 for the inland topography and multibeam bathymetry for 2011. The composite data set has some gaps along the coast, where water depth is too shallow for the vessels operating the multibeam to obtain good quality data but still beyond the reach of the LIDAR. A seamless topo-bathymetric-DTM was produced by interpolation after resampling to a 5 m grid. For the interpolation, we used the SAGA-Close Gaps function with a tension threshold of 0.1. The Close Gaps function uses a method commonly called minimum curvature under tension to interpolate missing data [38]. This DTM is then combined with the subsurface database to produce, first, a 3D geological model of the Quaternary sediments and, second, a thickness model. The 3D geological model was generated by combining the DTM, surface geological line-work and downhole borehole and geophysical data to enable the geologist to construct cross-sections by correlating boreholes and the outcrops to produce a geological fence diagram. Interpolation between the nodes along the drawn sections and the limits of the units produces a solid model comprising a stack of triangulated objects each corresponding to one of the geological units present. For the interpolation of the 3D geological model, we used GSI3D^{TM} software (Keyworth, UK, version 2013) [39] as described in [19]. Then, the elevation of the top and base of each geological unit (i.e., its thickness) is calculated by triangulating between digitized nodes along the cross-sections and nodes around the edges of unit coverages. These tops, bases and thicknesses are then exported from the model as grids with a user-defined cell size. Using empirical knowledge, supported by data from the BGS National Geotechnical Database [40], the litho-stratigraphical units expressed in the model were assigned average percentages for fines (<0.63 mm), sand (0.63–2 mm) and coarse material (>2 mm) (Table A1 [19]). The chalk consists mainly of silt particles and, although it forms a consolidated rock, was added to the fines fraction. All organic material was also assigned to the fines fraction. The geological model was queried to calculate the average grain size distribution of the geological sequence at each node of a 5 × 5 m grid. The resulting data were converted into raster thickness layers.

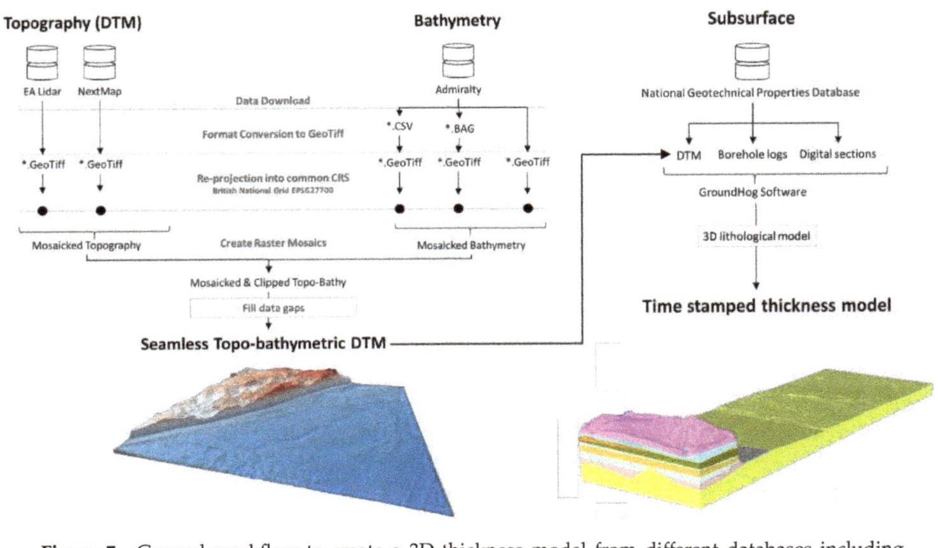

Figure 7. General workflow to create a 3D thickness model from different databases including topo-bathymetric and subsurface consolidated and unconsolidated sediment fractions.

3. Results

Figure 8 shows the different sediment fractions along the Happisburgh South transect for both the unconsolidated material (e.g., beach deposits) and consolidated platform. The sediment fractions (coarse, fine and sand) shown have been extracted from the raster outputs numbers 24, 25, 26 on Table A2 for the consolidated material and outputs numbers 66, 67 and 68 for the unconsolidated material at $t = 1$ day. QGIS (v 3.10.2) and the Profile-Tool (v 4.1.7) plugin were used to extract elevations along the transect. Figure 8a shows the sediment fractions (as thicknesses) for the unconsolidated material, which for Happisburgh represents the beach material. Given that the 3D grid is constructed at a 5 m interval, a thickness of 1 m is equivalent to 25 m^3 of material per spatial cell. For this transect, the beach deposit is dominated by sand with a relatively uniform thickness of 0.8 m along the beach (and a maximum of 1.34 m). The equivalent thickness of coarse material is an order of magnitude less (about 0.08 m with a maximum of 0.15 m). Fine material is negligible (<0.01 m). The unconsolidated material is also dominated by the sand fraction with a maximum thickness of 37 m, followed the fine fraction (maximum thickness of 10 m) and coarse material (always < 2.7 m). By adding together all fractions for the consolidated and unconsolidated material, we obtain the top elevation of the consolidated platform and actual topographic elevation, respectively (Figure 8c).

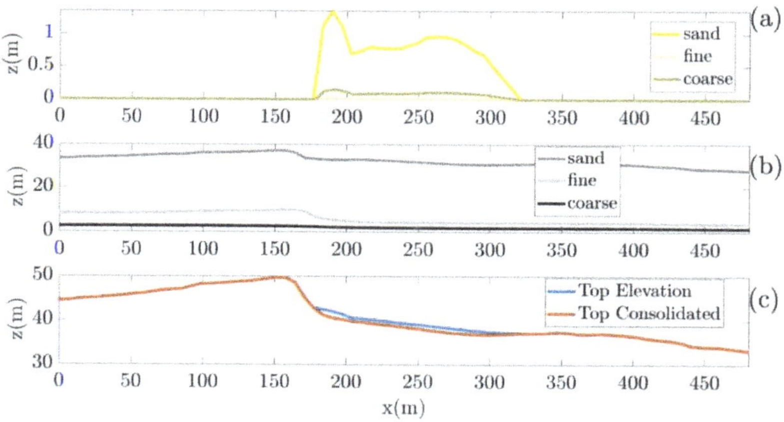

Figure 8. Visualisation of unconsolidated (**a**) and consolidated (**b**) sediment fractions (fine, sand and coarse) along Happisburgh South transect. Top elevations of the consolidated and total (i.e., consolidated + unconsolidated) elevation are shown in panel (**c**).

Figure 9a shows the beach deposit thickness, obtained from the difference between the topographic elevation and the consolidated platform top elevation, overlaid on an aerial image. Beach deposit thickness varies across and along the shoreline, ranging from 4 m in the north, minimal to non-existent along the undefended central section, and 2–3 m in the southern sector. As expected, the simulated lowering of the consolidated platform (Figure 9b) is largest at places where the beach deposits thickness is least and cannot offer protection. These differences in beach thickness, combined with the different defence structures close to each transect, have an appreciable effect on the simulated elevation changes (Figures 10–12).

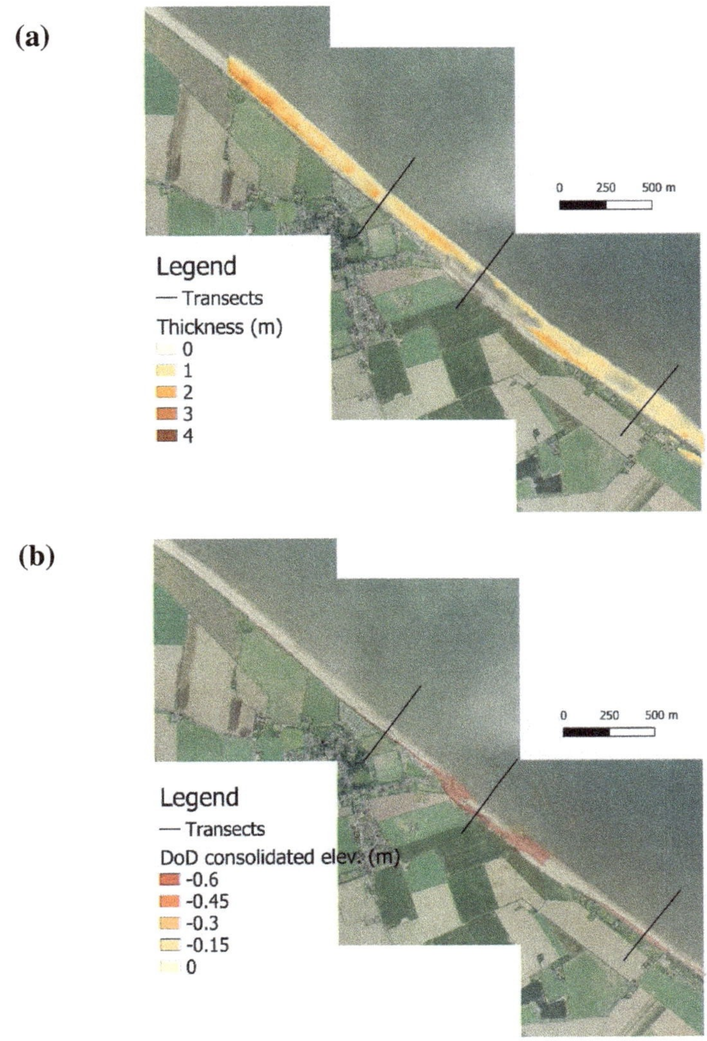

Figure 9. Beach deposit thickness (**a**) and simulated consolidated platform elevation change (**b**) along the study area. North, Central and South transects are shown as black solid lines. (source: Aerial photography© UKP/Getmapping Licence No. UKP2006/01).

Figure 10 shows the elevation change of the topographic elevation and consolidated elevation along the North transect computed for the 360-day run. The overall topographic elevation (Figure 10a) has been obtained by adding together all fractions for the consolidated and unconsolidated material of the raster files numbers 24, 25, 26 and 66, 67, 68 on Table A2. The top elevation of the consolidated platform (Figure 10b) has been obtained by adding all fractions of the consolidated material (i.e., outputs 24, 25, 26 on Table A2). On the North transect, where beach thickness is low (0–1 m) the topographic elevation and the upper elevation of the consolidated platform coincide. This section of coast is protected by sheet piling, which significantly reduces the erosion behind the defences. As expected, the simulated elevation change is concentrated at the toe of the cliff, which retreats 6 m landward over the year. Elevation decreases along the entire profile (i.e., no accretion occurs) with a maximum vertical erosion of 1 m.

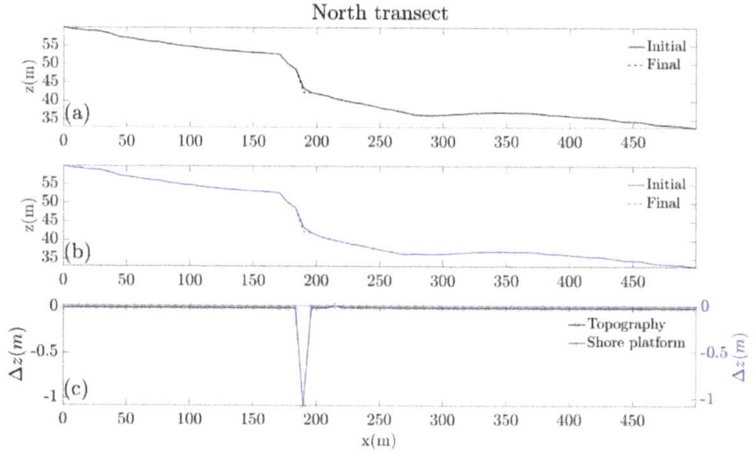

Figure 10. Visualisation of topography and consolidated elevation change along Happisburgh North transect; panel (**a**) and (**b**) show the initial and final topography and consolidated elevation respectively. Panel (**c**) shows the elevation change (e.g., final minus initial elevation) for both topography (black line and circles) and consolidated platform (blue solid line).

Figure 11 shows the elevation change of the top total elevation and consolidated elevation along the Central transect. As with the North transect, where beach thickness is low (0–1 m), the topographic elevation coincides with the top of the consolidated material. As expected for this undefended stretch of coast, cliff toe retreat and vertical erosion are more rapid than that simulated for the defended North and South transects. Cliff toe retreat is approximately 33 m and vertical erosion is up to 5 m. The beach deposit is decreased by a maximum of 0.7 m but remains thick enough to protect the consolidated platform underneath, the elevation of which remains unchanged.

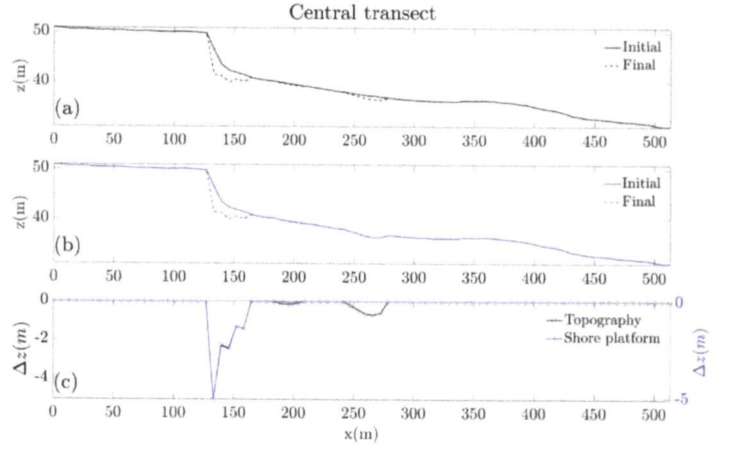

Figure 11. Visualisation of topography and consolidated elevation change along Happisburgh Central transect; panel (**a**) and (**b**) shows the initial and final topography and consolidated elevation, respectively. Panel (**c**) shows the elevation change (e.g., final minus initial elevation) for both topography (black line and circles) and consolidated platform (blue solid line).

Figure 12 shows the equivalent results along the South transect, where the beach thickness is greater (1–2 m). Cliff toe horizontal location remains unchanged because this section is protected by a sea-wall. Topographic lowering is no longer restricted to the proximity of the cliff toe and is appreciable, with an average value of −0.8 m, along the entire active transect with a maximum vertical erosion of −2 m close to the sea-wall toe. However, lowering of the consolidated platform is still constrained to the region nearby the sea-wall toe where it reaches a maximum of 2 m. This implies that the topographic lowering is mostly due to vertical erosion of the consolidated platform near the sea-wall toe while, seawards, the lowering is due to beach material being lost due to longshore sediment transport gradient. Even here, the beach appears to remain thick enough to protect the underlying consolidated platform from lowering.

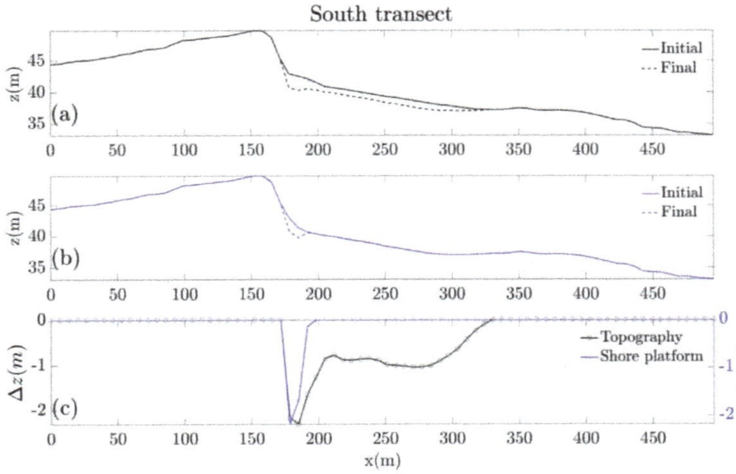

Figure 12. Visualisation of topography and consolidated elevation change along Happisburgh South transect; panel (**a**) and (**b**) shows the initial and final topography and consolidated elevation, respectively. Panel (**c**) shows the elevation change (e.g., final minus initial elevation) for both topography (black line and circles) and consolidated platform (blue solid line).

Figure 13 shows the initial and final simulated cliff top lines along with a Digital Elevation Model (DEM) created from the topographic elevation. Cliff top lines were extracted using the CliffMetric automatic delineation algorithm proposed by Payo et al. (2015). Cliff top and toe location points have been converted into lines and the shortest distance between the final and initial cliff top points used as an indication of cliff retreat. A maximum cliff top retreat of 30 m is obtained in the north of the non-defended coastal section, while the cliff top remains unchanged (i.e., differences smaller than one diagonal cell length or 7 m) for the defended sections. The lines are plotted on a 2D view of the study area and 3D view with aerial imagery as background to facilitate geolocation.

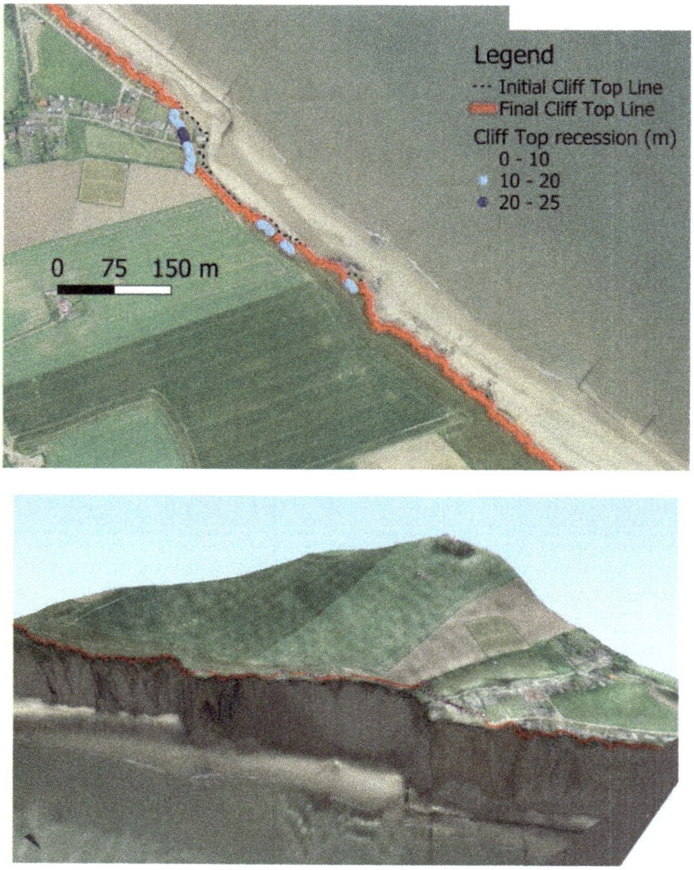

Figure 13. 2D (**top**) and 3D (**bottom**) views of the initial and final cliff top lines obtained from the simulated DEM changes. Cliff top recession larger than one diagonal cell length is shown as coloured circles. (source: Aerial photography© UKP/Getmapping Licence No. UKP2006/01)

4. Discussion

The credibility of any simulation of mesoscale coastal geomorphological change, for both specialist and non-specialists, depend on three essentials: (1) Data—the quality of the data on which the simulation model is based; (2) Model—the fidelity of the algorithms, the validity of the assumptions and the correctness of the underlying computer code; and (3) Visualisation—representation of model outputs that can be clearly linked to the underlying model prediction and the real-world problem. These aspects are considered further here in relation to the Happisburgh case.

Regarding the quality of the data, we have shown how the Happisburgh study zone can be represented using the CoastalME data structure (Figure 6) by combining existing digital elevation and sub-surface data (Figure 7) to initialise a 3D thickness model. This 3D thickness model is based on different types of data (LiDAR-DTM + Multibeam bathymetry + borehole logs and cross-shore sections of the subsurface) and the steps required to combine this variable set of data types are transparent and reproducible. In particular, the discrete nature of the databases (i.e., boreholes logs and cross-sections) and the data gap of bathymetry in the very shallow nearshore region and offshore region can be filled-in using different interpolation approaches. Combining these three databases allows a better representation of the supply of transportable material, which is essential to obtain an accurate estimate of the nearshore sediment balance and ultimately the coastal morphological change. The amount of transportable material can be visualised in a number of ways; as thicknessses of different sediment fractions along selected transects (Figure 8) or as a raster map of beach thickness (Figure 9).

The fidelity of the algorithms included in the CoastalME model composition used for the Happisburgh digital model was discussed in detail by [19]. They showed that the main features (i.e., cliff, beach, shore-platform, sea-wall, and palisade) and its interaction were incorporated in the model composition (i.e., structural validity). Here, we have illustrated the validity of the assumptions and the competence of the code by showing how the digital model forced with a one-year simulation of wave and tidal forcing shows a rich and plausible behaviour as supported from field data [20] and expected for an eroding soft coastline with defended and non-defended coastal stretches (Figures 10–12). Furthermore, the simulation outputs also provide insights of a potentially significant platform down-wearing in-front of the sea-wall that is very difficult to observe in the real system since the platform in-front of the sea-wall is often covered by the beach, precluding direct observation of the lowering (Figure 9).

We have shown how model outputs can be clearly linked to the underlying model prediction as 1D, 2D and 3D geo-indicators. Contrary to the loose linkage presented by [12] where 1D model outputs were translated into 3D-DEM during the visualisation stage, we have shown here how the CoastalME model output "sediment_top_elevation", representing the top total sediment elevation, is effectively a georeferenced evolving 3D-DEM. This sidesteps the need for further assumptions during the visualisation stage (i.e., cliff toe line being parallel to cliff top line as done by [12]) that might not be always scientifically justified. We have shown how other 1D and 2D geo-indicators of change used by specialist and non-specialist [41] such as sediment fraction composition (Figure 8), elevation profile (Figures 10–12), and cliff top line (Figure 13) can be extracted from the evolving 3D-DEM. CoastalME uses the georeferenced system used for the 3D-thickness input model as the default reference system for all the vector and raster outputs (Table A2), which makes it very straightforward to combine the model outputs with other georeferenced information that facilitates communication of the outputs. For example, we have shown how the outputs can be combined in 2D and 3D with aerial imagery, and vector lines (Figure 13).

We acknowledge several key limitations that presently hinder the more widespread use of the proposed methodology (Figure 7) to create digital representation at other locations throughout the UK. In no particular order, these limitations are;

1. **Lack of a national coastal defence database**. The Environment Agency's Asset Information Management System contains the location of flood defences owned, managed or inspected by the EA and coastal protection assets managed by other operating authorities. Data includes defence type (i.e., groin, sheet pile, palisade, etc.) location and main dimensions as designed and may include a condition grade from an asset inspection [42]. However, not all attributes are present. Additionally, private defensive structures are excluded. This lack of data makes it very difficult to ensure that the coastal interventions represented in the model correspond with the coastal defence on the ground. For the example of Happisburgh presented here, the depth of the sheet piles is unknown making it impossible to assess the risk of scouring undermining the sheet pile and consequent ultimate failure.
2. **Need for more frequently updated topo-bathymetric databases.** The coastal topography and bathymetry is dynamic and continuously changing over time. The EA-LiDAR DTM and UKHO multi-beam bathymetries have good spatial coverage but provide only snap-shots at given dates of the state of the physical system. As the different agencies in charge of updating the DTMs operate independently and with different budgets, the date of the most up to date DTM available might vary from place to place. While daily updates of the topo-bathymetry DTM are unlikely to be needed for the purpose of exploring "what if" scenarios at decadal and longer time scales, they are extremely valuable for ongoing model validation.
3. **Sensitivity of simulation outputs to interpolations and modeller assumptions.** Due to the discrete nature of geotechnical data and the existence of gaps in topographic and bathymetric

data, interpolation will likely remain an important part of any simulation model. The choice of model resolution (spatial and temporal) is one of the many decisions that can affect the simulation of sub-mesoscale scale features. Sub-grid features (and processes) are necessarily smoothed out by interpolation onto coarser grids, and this may influence the depiction and prediction of mesoscale morphological change. Although it has been argued [43,44] that mesoscale coastal morphodynamics is substantially decoupled from small-scale processes, this is clearly an aspect of model development that requires careful attention. Sensitivity testing of the overall simulation outputs to different interpolation, resolution and other model assumptions ideally require a standardised approach.

4. **The need for a curator of model composition and model instances**. To realise maximum benefit from the resource investment in environmental/earth science models, it is necessary to record a rich set of model metadata. This metadata should include attributes such as which environmental/earth science discipline is involved, and which parameters are input and output in the modelling process. In 2016, NERC created the Model Metadata Application (http://model-search.nerc.ac.uk/) to help users discover and locate the existence of models, and also descriptive or "usage" metadata which is of relevance when making use of a model, for example, when using a model code developed by another researcher. As coastal model compositions and coastal model instances become available in the future, they will need to be recorded accordingly in the Model Metadata Application or any similar platform.

5. Conclusions

It has been recognised by [12] that by linking an erosion model with a GIS and then developing the resulting spatial information into visualisations of the evolving coastal environment, information on the changing hazard of future coastal recession can be made more accessible to non-specialists. Non-specialists generally seem to welcome the additional detail and realism that the visualisations provide, and acknowledge the communication and awareness-raising value of the images. The credibility of the resulting virtual future landscapes is also enhanced by their derivation from scientific data provided by specialists using simulation models. A key issue that emerges is that the quest for realism in visualisation can lead to more detailed data demands than can be provided by the underlying scientific model supplying the rationale for the landscape predictions.

In this work we have attempted to address a knowledge gap concerning the important but still unexplored challenge of communicating simulations of mesoscale coastal dynamic evolution to specialist and non-specialist audiences. We have shown how the risk of simulation model outcomes becoming loosely connected to more realistic visualisations of model outcomes can be minimised by using the Coastal Modelling Environment framework. CoastalME is a bespoke modelling framework for coastal mesoscale morphological modelling with close linkages between the scientific model abstractions, in the form of lines, areas and volumes, the 3D representation of the coastal topo-bathymetry elevation and shallow sub-surface sediment fraction composition. We have proposed and illustrated through the study case of Happisburgh, East England (UK), a transparent and reproducible methodology to merge the required variety of data types and formats into a 3D-thickness model that is used to initialise the simulation. We also highlight some of the limitations that are preventing the direct adoption of the methodology proposed.

Author Contributions: Conceptualization: A.P., M.W., M.A.E., J.R.F. & J.S.; methodology, A.P.; software, A.P. & M.W.; formal analysis, A.P.; resources, M.A.E.; data curation, A.P.; writing—original draft preparation, A.P.; writing—review and editing, J.R.F., J.S., M.A.E., M.W.; visualization, A.P. & M.W.; supervision, M.A.E.; project administration, A.P.; funding acquisition, M.A.E. All authors have read and agreed to the published version of the manuscript.

Funding: This work was funded by NERC (http://www.nerc.ac.uk) as part of the BLUEcoast project (https://projects.noc.ac.uk/bluecoast/) (NE/N015649/1).

Acknowledgments: Andres Payo and Michael A. Ellis publish with the permission of the Executive Director, British Geological Survey (UKRI).

Conflicts of Interest: The authors declare no conflict of interest.

Appendix A

Table A1. CoastalME composition model inputs for Happisburgh one year simulation (after [19]).

Input	Value
Required for a generic landscape evolution model	
Run duration	360 days
Time step	1 h
Wave heights, direction, period	UKCP09 hindcast data
Topo and bathymetric Digital Elevation Model	LiDAR year 1999 & Multibeam 2011
Tides	Reconstruction of tidal signal using Cromer tide gauge data from 1999 to 2017
Residual elevation	Difference of Cromer tide gauge elevation and tidal levels (gap filled assuming residuals follow a normal distribution)
CoastalME Datum	+40 m above basement level
Coarse, sand and fine sediment content	BGS thickness model
Coarse, sand and fine availability factor	0.3; 0.7; 1.0
Boundary conditions	Open boundaries (i.e., sediment at the boundaries is allotted to exit the grid but no external sediment inputs are assumed over the simulated period)
Required for COVE-sediment sharing module	
CERC coefficient	0.79
Length of normal profiles used to create the polygons	800 m
Required for CSHORE-wave propagation module	
Breaker ratio parameter γ	0.8
Friction factor f_b	0.015
Required for SCAPE-beach & platform interaction	
Rock strength and hydrodynamic constant, R	$R_{\text{Platform}} = 8 \times 10^4$ [$m^{9/4}s^{2/3}$] $R_{\text{Cliff}} = 8 \times 10^2$ [$m^{9/4}s^{2/3}$]
Beach volume & and beach thickness	Derived from BGS thickness model

Full list of parameters provided in Table S1 in [19].

Table A2. CoastalME GIS simulation outcomes name and type. Sorted in filename alphabetical order.

ID	Output name *	Type
1	active_zone	Raster
2	actual_beach_erosion	Raster
3	avg_sea_depth	Raster
4	avg_susp_sed	Raster
5	avg_wave_angle	Vector
6	avg_wave_height	Raster
7	avg_wave_orientation	Raster

Table A2. *Cont.*

ID	Output name *	Type
8	basement_elevation	Raster
9	beach_change_net	CSV
10	beach_deposition	CSV
11	beach_deposition	Raster
12	beach_erosion	CSV
13	beach_mask	Raster
14	beach_protection	Raster
15	breaking_wave_height	Vector
16	cliff_collapse	Raster
17	cliff_collapse_deposition	CSV
18	cliff_collapse_deposition	Raster
19	cliff_collapse_erosion	CSV
20	cliff_collapse_net	CSV
21	cliff_notch	Vector
22	coast	Vector
23	coast_curvature	Vector
24	cons_sed_coarse_layer_X	Raster
25	cons_sed_fine_layer_X	Raster
26	cons_sed_sand_layer_X	Raster
27	deep_water_wave_angle	Vector
28	deep_water_wave_height	Raster
29	deep_water_wave_orientation	Raster
30	downdrift_boundary	Vector
31	ErosionPotential	CSV
32	intervention_class	Raster
33	intervention_height	Raster
34	invalid_normals	Vector
35	landform_class	TIF
36	local_cons_sediment_slope	Raster
37	mean_wave_energy	Vector
38	node	Vector
39	normals	Vector
40	platform_erosion	CSV
41	polygon	Vector
42	polygon_gain_or_loss	Raster
43	polygon_raster	Raster
44	polygon_updrift_or_downdrift	Raster
45	potential_beach_erosion	Raster

Table A2. *Cont.*

ID	Output name *	Type
46	potential_platform_erosion	Raster
47	rcoast	Raster
48	rcoast_normal	Raster
49	sea_area	CSV
50	sea_depth	Raster
51	sediment_top_elevation	Raster
52	shadow_boundary	Vector
53	shadow_downdrift_zones	Raster
54	shadow_zones	Raster
55	still_water_level	CSV
56	susp_sed	Raster
57	suspended_sediment	CSV
58	top_elevation	Raster
59	total_actual_beach_erosion	Raster
60	total_actual_platform_erosion	Raster
61	total_beach_deposition	Raster
62	total_cliff_collapse	Raster
63	total_cliff_collapse_deposition	Raster
64	total_potential_beach_erosion	Raster
65	total_potential_platform_erosion	Raster
66	uncons_sed_coarse_layer_X	Raster
67	uncons_sed_fine_layer_X	Raster
68	uncons_sed_sand_layer_X	Raster
69	wave_angle	Vector
70	wave_energy	Vector
71	wave_height	Raster
72	wave_orientation	Raster
73	wave_period	Raster

* The output file name is completed by a four digit number indicating the time interval (i.e., wave_period_0001 is the output for interval 1).

References

1. French, J.R.; Burningham, H. Coastal geomorphology: Trends and challenges. *Prog. Phys. Geogr. Earth Environ.* **2009**, *33*, 117–129. [CrossRef]
2. French, J.; Payo, A.; Murray, B.; Orford, J.; Eliot, M.; Cowell, P. Appropriate complexity for the prediction of coastal and estuarine geomorphic behaviour at decadal to centennial scales. *Geomorphology* **2016**, *256*, 3–16. [CrossRef]
3. Kamphuis, J. Beyond the limits of coastal engineering. In *Coastal Engineering 2006*; World Scientific: Singapore, 2007; Volumes 5, pp. 1938–1950.
4. Hanson, H.; Aarninkhof, S.; Capobianco, M.; Jimenez, J.; Larson, M.; Nicholls, R.; Plant, N.; Southgate, H.; Steetzel, H.; Stive, M. Modelling of coastal evolution on yearly to decadal time scales. *J. Coast. Res.* **2003**, *19*, 790–811.

5. Bray, M.; Hooke, J.; Carter, D. Planning for sea-level rise on the south coast of england: Advising the decision-makers. *Trans. Inst. Br. Geogr.* **1997**, *22*, 13–30.
6. Green, D.R. The role of public participatory geographical information systems (PPGIS) in coastal decision-making processes: An example from Scotland, UK. *Ocean Coast. Manag.* **2010**, *53*, 816–821. [CrossRef]
7. Blunkell, C.T. Local participation in coastal adaptation decisions in the UK: Between promise and reality. *Local Environ.* **2017**, *22*, 492–507. [CrossRef]
8. Robinson, S. Conceptual modelling for simulation part II: A framework for conceptual modelling. *J. Oper. Res. Soc.* **2008**, *59*, 291–304. [CrossRef]
9. French, J.; Burningham, H.; Thornhill, G.; Whitehouse, R.; Nicholls, R.J. Conceptualising and mapping coupled estuary, coast and inner shelf sediment systems. *Geomorphology* **2016**, *256*, 17–35. [CrossRef]
10. Payo, A.; Hall, J.W.; French, J.; Sutherland, J.; van Maanen, B.; Nicholls, R.J.; Reeve, D.E. Causal loop analysis of coastal geomorphological systems. *Geomorphology* **2016**, *256*, 36–48. [CrossRef]
11. Forrester, J.W.; Senge, P.M. *Tests for Building Confidence in System Dynamics Models*; System Dynamics Group, Sloan School of Management, Massachusetts Institute of Technology Cambridge: Cambridge, MA, USA, 1978.
12. Brown, I.; Jude, S.; Koukoulas, S.; Nicholls, R.; Dickson, M.; Walkden, M. Dynamic simulation and visualisation of coastal erosion. *Comput. Environ. Urban Syst.* **2006**, *30*, 840–860. [CrossRef]
13. Appleton, K.; Lovett, A. Gis-based visualisation of rural landscapes: Defining 'sufficient'realism for environmental decision-making. *Landsc. Urban Plan.* **2003**, *65*, 117–131. [CrossRef]
14. Cowell, P.J.; Roy, P.S.; Jones, R.A. Shoreface translation model: Computer simulation of coastal-sand-body response to sea level rise. *Math. Comput. Simul.* **1992**, *33*, 603–608. [CrossRef]
15. Walkden, M.J.; Hall, J.W. A mesoscale predictive model of the evolution and management of a soft-rock coast. *J. Coast. Res.* **2011**, *27*, 529–543. [CrossRef]
16. James, J.; Region, A.; House, K.; Way, G.; Goldhay, O. Sediment input from coastal cliff erosion. *Br. Geol. Surv.* **1995**, *74* (TR/577/4/A).
17. Dickson, M.E.; Walkden, M.J.A.; Hall, J.W. Systemic impacts of climate change on an eroding coastal region over the twenty-first century. *Clim. Chang.* **2007**, *84*, 141–166. [CrossRef]
18. Walkden, M. *Scape Modelling of Shore Evolution: Cromer to Cart Gap, Appendix C of the Cromer to Winterton Ness Coastal Management Study*; Royal Haskoning Report for Mott Macdonald, on behalf of North Norfolk District Council: Penryn, UK, July 2013.
19. Payo, A.; Walkden, M.; Ellis, M.; Barkwith, A.; Favis-Mortlock, D.; Kessler, H.; Wood, B.; Burke, H.; Lee, J. A quantitative assessment of the annual contribution of platform downwearing to beach sediment budget: Happisburgh, England, UK. *J. Mar. Sci. Eng.* **2018**, *6*, 113. [CrossRef]
20. López, P.M.; Payo, A.; Ellis, M.A.; Criado-Aldeanueva, F.; Jenkins, G.O. A method to extract measurable indicators of coastal cliff erosion from topographical cliff and beach profiles: Application to North Norfolk and suffolk, East England, UK. *J. Mar. Sci. Eng.* **2020**, *8*, 20.
21. Payo, A.; Walkden, M.; Barkwith, A.; Ellis, A.M. Modelling rapid coastal catch-up after defence removal along the soft cliff coast of Happisburgh, UK. In Proceedings of the 36th International Conference on Coastal Engineering, Baltimore, MD, USA, 30 July–3 August 2018; ASCE: Baltimore, MD, USA; p. 2.
22. Walkden, M.; Watson, G.; Johnson, A.; Heron, E.; Tarrant, O. *Coastal Catch-Up Following Defence Removal at Happisburgh*; Coastal Management, ICE: London, UK, 2016; pp. 523–532.
23. Brown, S.; Barton, M.E.; Nicholls, R.J. Shoreline response of eroding soft cliffs due to hard defences. In *Proceedings of the Institution of Civil Engineers-Maritime Engineering*; Thomas Telford Ltd.: London, UK, 2014; Volume 167, pp. 3–14.
24. Clayton, K.M. Sediment input from the norfolk cliffs, Eastern England—A century of coast protection and its effect. *J. Coast. Res.* **1989**, *5*, 433–442.
25. Poulton, C.V.; Lee, J.; Hobbs, P.; Jones, L.; Hall, M. Preliminary investigation into monitoring coastal erosion using terrestrial laser scanning: Case study at happisburgh, norfolk. *Bull. Geol. Soc. Norfolk* **2006**, *56*, 45–64.
26. Frew, P. Adapting to coastal change in North Norfolk, UK. In *Proceedings of the Institution of Civil Engineers-Maritime Engineering*; Thomas Telford Ltd.: London, UK, 2012; Volume 165, pp. 131–138.
27. Hayman, S. *Eccles to Winterton on Sea Coastal Defences*; SH/RG BF190712; Environment Agency, Norfolk Broads Forum: Norfolk, UK, 2012; p. 3.

28. Hobbs, P.; Pennington, C.; Pearson, S.; Jones, L.; Foster, C.; Lee, J.; Gibson, A. *Slope Dynamics Project Report: Norfolk Coast (2000–2006)*; British Geological Survey: Nottingham, UK, 2008; p. 166.
29. Lunkka, J.P. Sedimentation and iithostratigraphy of the north sea drift and lowestoft till formations in the coastal cliffs of northeast norfolk, england. *J. Quat. Sci.* **1994**, *9*, 209–233. [CrossRef]
30. Lee, J.R. Early and Middle Pleistocene Lithostratigraphy and Palaeo-Environments in Northern East Anglia. Ph.D. Thesis, Royal Holloway, Egham, Surrey, University of London, London, UK, 2003.
31. Lee, J.R.; Phillips, E.R. Progressive soft sediment deformation within a subglacial shear zone-a hybrid mosaic-pervasive deformation model for middle pleistocene glaciotectonised sediments from eastern england. *Quat. Sci. Rev.* **2008**, *27*, 1350–1362. [CrossRef]
32. Lee, J.R.; Phillips, E.; Rose, J.; Vaughan-Hirsch, D. The middle pleistocene glacial evolution of northern east anglia, uk: A dynamic tectonostratigraphic–parasequence approach. *J. Quat. Sci.* **2017**, *32*, 231–260. [CrossRef]
33. Parfitt, S.A.; Ashton, N.M.; Lewis, S.G.; Abel, R.L.; Coope, G.R.; Field, M.H.; Gale, R.; Hoare, P.G.; Larkin, N.R.; Lewis, M.D. Early pleistocene human occupation at the edge of the boreal zone in northwest Europe. *Nature* **2010**, *466*, 229. [CrossRef] [PubMed]
34. West, R.G. *The Pre-Glacial Pleistocene of the Norfolk and Suffolk Coasts: With Contrib. By Pep Norton*; Cambridge University Press: Cambridge, UK, 1980.
35. Payo, A.; Favis-Mortlock, D.; Dickson, M.; Hall, J.W.; Hurst, M.D.; Walkden, M.J.A.; Townend, I.; Ives, M.C.; Nicholls, R.J.; Ellis, M.A. Coastal modelling environment version 1.0: A framework for integrating landform-specific component models in order to simulate decadal to centennial morphological changes on complex coasts. *Geosci. Model Dev.* **2017**, *10*, 2715–2740. [CrossRef]
36. van Maanen, B.; Nicholls, R.J.; French, J.R.; Barkwith, A.; Bonaldo, D.; Burningham, H.; Brad Murray, A.; Payo, A.; Sutherland, J.; Thornhill, G.; et al. Simulating mesoscale coastal evolution for decadal coastal management: A new framework integrating multiple, complementary modelling approaches. *Geomorphology* **2015**, *256*, 68–80. [CrossRef]
37. Payo, A.; Favis-Mortlock, D.; Dickson, M.; Hall, J.W.; Hurst, M.; Walkden, M.J.A.; Townend, I.; Ives, M.C.; Nicholls, R.J.; Ellis, M.A. Coastalme version 1.0: A coastal modelling environment for simulating decadal to centennial morphological changes. *Geosci. Model Dev. Discuss.* **2016**, *2016*, 1–45.
38. Smith, W.; Wessel, P. Gridding with continuous curvature splines in tension. *Geophysics* **1990**, *55*, 293–305. [CrossRef]
39. Kessler, H.; Mathers, S.; Sobisch, H.-G. The capture and dissemination of integrated 3d geospatial knowledge at the british geological survey using gsi3d software and methodology. *Comput. Geosci.* **2009**, *35*, 1311–1321. [CrossRef]
40. Self, S.; Entwisle, D.; Northmore, K. *The Structure and Operation of the BGS National Geotechnical Properties Database. Version 2*; British Geological Survey: Nottingham, UK, 2012; p. 68.
41. Carapuço, M.M.; Taborda, R.; Silveira, T.M.; Psuty, N.P.; Andrade, C.; Freitas, M.C. Coastal geoindicators: Towards the establishment of a common framework for sandy coastal environments. *Earth Sci. Rev.* **2016**, *154*, 183–190.
42. Flikweert, J.; Lawton, P.; Roca-Collel, M.; Simm, J. *Guidance on Determining Asset Deterioration and the Use of Condition Grade Deterioration Curves*; Environment Agency: Bristol, UK, 2009.
43. Lazarus, E.; Ashton, A.; Murray, A.B.; Tebbens, S.; Burroughs, S. Cumulative versus transient shoreline change: Dependencies on temporal and spatial scale. *J. Geophys. Res. Earth Surf.* **2011**, *116*. [CrossRef]
44. Murray, A.B.; Coco, G.; Goldstein, E.B. Cause and effect in geomorphic systems: Complex systems perspectives. *Geomorphology* **2014**, *214*, 1–9. [CrossRef]

Journal of
Marine Science and Engineering

MDPI

Case Report

Effects of Repeated Sand Replenishment Projects on Runs of a Beach-Spawning Fish, the California Grunion

Karen L. M. Martin [1,*] and Loni C. Adams [2]

1 Department of Biology, Pepperdine University, Malibu, CA 90236, USA
2 California Department of Fish and Wildlife, Marine Region, San Diego, CA 92123, USA; Loni.Adams@wildlife.ca.gov
* Correspondence: karen.martin@pepperdine.edu

Received: 29 December 2019; Accepted: 3 March 2020; Published: 6 March 2020

check for updates

Abstract: Beach habitats are diminishing globally, particularly in urban areas, as sea-level rise, erosion, and shoreline hardening, along with reduced sediment inputs, combine to squeeze the coast. In California, USA an endemic marine fish, the California grunion, spawns on sandy beaches during late-night spring tides. Its unique recreational fishery is managed by the California Department of Fish and Wildlife. The City of Oceanside, CA contracts for annual harbor dredging and, after testing, places the sandy sediment on its public beach. The effects on local beach wildlife from this annual sand replenishment are not known. We examined the effect of this repeated activity as a case study over three years on the spawning runs of the California grunion. Some spawning runs occurred in all three years, but the fish avoided areas with high scarps in the intertidal zone that developed following sand placement activity. Grunion spawning runs have declined in the habitat range as a whole over the past two decades, and those in Oceanside have declined to an even greater extent. Increasing sandy beach habitat can be beneficial to wildlife, but the method of placement, timing of the project, and fate of the beach afterward can modulate or prevent beneficial effects. Frequent repetition of sand placement may accumulate impacts without allowing sufficient time for the ecosystem to recover. Rather than improving the habitat, these repeated projects in Oceanside may degrade the spawning habitat for the grunion. Alternative discharge methods and locations, slope and elevation designs, sediment volumes, and greater care in beach fill practices should be implemented to reduce future impacts.

Keywords: beach nourishment; beach restoration; ecosystem management; substrate; reproductive habitat; human impacts; beach-spawning fishes; essential fish habitat

1. Introduction

Sand replenishment on beaches is commonly used as a means of restoring, building up elevation or expanding beach width [1,2]. It is also used as a beneficial use of discarding the materials of a dredge operation for harbor navigation. The slope, timing, frequency, amount and type of materials placed on the beach are all tested and managed, as all may affect the biota that either live there permanently or use the beach ecosystem for some part of their life cycle [3–6]. The effects of beach nourishment are poorly known for most beach biota, either in the shorter- or longer-term [7–9]. Knowing how coastal ecosystems and beaches and their sediments respond to storm conditions and repeated beach replenishments is a pressing question for resource managers, particularly because beach replenishment is now the preferred option in the United States for short-term stabilization of any eroding coastline which is used for valued recreation or tourism [10].

Oceanside Harbor in San Diego County, California, USA is a marina for commercial and recreational vessels and is also used by the United States Marine Base at Camp Pendleton. To maintain navigability for large vessels, the harbor is dredged on an annual basis under the auspices of the US Army Corps of Engineers (USACE). The harbor dredge materials are tested for grain size and contaminants. If found to be within suitable limits, they are placed on the shoreline on Oceanside City Beach nearby as a form of sand replenishment or "beach nourishment" to increase the width of a recreational beach adjacent to an urban shoreline.

Sand replenishment projects are often touted as being beneficial to beach wildlife because in theory, they create more habitat by the placement of substrate [2,5]. "Beach restoration" is a term often used for sand replenishment projects, but this "restoration" generally does not include plantings or re-introduction of native species, whether adults or propagules, as is typical for other types of ecological restorations [11–13]. The timing and frequency of projects also influence ecological recovery.

California is home to many animals that use sandy beaches as nursery areas, including birds, mammals, reptiles, and fishes. The California grunion *Leuresthes tenuis* (Atherinopsidae) is a beach-spawning marine fish that is famous for emerging fully from the water during spawning runs under a full or new moon [14,15]. They bury their eggs under 10–20 cm of sand, where the eggs remain out of the water above the water line, until being washed free nearly two weeks later by rising tides. The spawning season may start as early as February and go through to August, with April through to June as the months of peak activity [16]. Runs can be forecast according to the tides and are most likely to occur within four days after a full or new moon, in a two-hour window following each nightly high tide [17].

The main habitat range of the California grunion is a few hundred kilometers along one of the most populated coasts in the world [18,19]. The city of Oceanside is in the heart of the southern California range for this endemic species. The sandy beaches of Oceanside and nearby towns are historically known to host significant grunion runs, to the extent that they are touted by the tourist literature as a visitor activity [20].

Most coastal construction projects in California must avoid disturbance to sandy beaches during key reproductive periods of beach-spawning birds and the California grunion [21,22]. Biological monitors may be required on beaches prior to, and for the duration of the project, and projects may be moved or temporarily halted if some species are present in the construction site.

Beach placement of harbor dredge material has taken place in Oceanside City Beach on nearly an annual basis since at least 1965. The permitting of this project is managed by the US Army Corps of Engineers and the California Coastal Commission. This annual dredging and beach replenishment project has in the past been completed by April, or before the peak spawning season of the California grunion. Recently, these projects have been permitted even during the peak spawning season for the California grunion. Sand replenishment projects were approved to take place during the grunion spawning season for Oceanside City beach in 2016, 2017 and 2018.

Although there have been dozens of sand replenishment projects along the coast of California over the past decades, the effects of beach sand replenishment on the California grunion have not been studied. For the California grunion, potential impacts could include disturbance to or stoppage of a spawning run during active beach filling, because of the disturbance of bright lights and noise [23]. Disturbance of the sand by vehicles during nest incubation and excessive burial of eggs by longshore drift after sand placement could prevent grunion hatchling emergence [3]. High turbidity of nearshore waters could alter survival during larval emergence, and changes to the substrate composition could affect the ability of the adult female fish to burrow into the sand for oviposition, or the eggs to wash free for hatching. How the California grunion chooses a beach for spawning is not fully understood, but changes in the substrate or sediment grain size may alter beach slope, and turbidity and sediment plumes may alter chemical signatures or strength and direction of waves that may assist in natal homing [24].

California grunion are difficult to monitor; they cannot be assessed by traditional fisheries methods such as trawl surveys or fishing reports, as they avoid nets and do not take a hook. Their legal recreational capture occurs only while on the shore during a bare-handed fishery [16,25]. Reporting a catch of California grunion is not required, and although individuals are required to have a California fishing license to hunt the grunion, observers suggest that many do not [25,26]. The most reliable method to monitor for grunion is to watch for spawning runs during the time of semilunar high tides [22,25].

Previous projects have not been required to monitor for grunion after the completion of sand placement. We hypothesized that grunion runs could be negatively affected by sand replenishment activities on the shoreline. We also hypothesized that grunion populations may be negatively impacted by frequent disturbances of repeated projects in Oceanside near to or during their spawning season.

2. Materials and Methods

The City of Oceanside has two sandy public beaches: Oceanside City Beach, in front of a residential area south of the San Luis Rey River mouth, and Oceanside Harbor Beach, in front of a marina north of the river mouth (Figure 1). Prior to the harbor jetties being built, the beaches south of San Luis Rey River were naturally wide beaches.

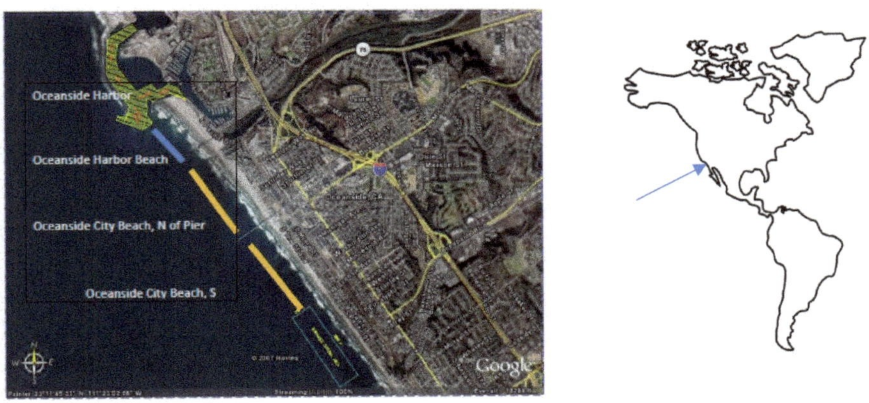

(a) (b)

Figure 1. (**a**), Camp Pendleton is north and east of Oceanside, just above the map outline. Oceanside Harbor Beach (blue line) is upcoast of the San Luis Rey River and outside of the filling zones. On Oceanside City Beach, for grunion spawning observation sites, Area 1 (yellow line) is north of the pier and Area 2 (yellow line) is south of the pier. Discharge pipe locations for 2016–2018 were within Areas 1 and 2. (**b**) Oceanside is on the West Coast of North America, shown by the blue arrow.

In the late 19th century, Oceanside's beach was approximately 100 m wide and used as a thoroughfare for horse-drawn wagons [27]. In Oceanside, CA, the prevailing ocean current direction is to the south. Long-shore sand transport typically moves sand slowly to replenish down-coast beaches [28], and tidal cycles affect beach profiles as well [29].

In 1942, the marine base Camp Joseph Pendleton was created, along with a jetty and a boat basin. However, the harbor soon silted up with sediments that previously maintained the width of Oceanside beaches [27]. After dredging the harbor to maintain its function, between 1942 and 1980 over 10 million cubic yards of material, including cobble, were deposited on Oceanside beaches [27]. Cobble did not appear on Oceanside beaches before 1965 [27].

Oceanside Harbor Beach (33° 12′ 12.86″ N, 117° 23′ 34.08″ W), upcoast from the replenishment project, is a 1500 feet long sandy beach from San Luis Rey Jetty, to North Jetty (Figure 2). It is a relatively flat beach with all three intertidal zones present as well as a wide supratidal and dry beach area. There was no beach-filling or bulldozing here between 2016 and 2018, although the beach was regularly raked and groomed to remove litter and debris.

Figure 2. Photo of Oceanside Harbor beach (South end), on 19 March 2018. This beach was outside of the project area and did not change in profile or character during the course of this study. Photo by L. C. Adams,

Two sites on Oceanside City Beach were monitored for habitat and grunion spawning activity before, during and after beach fill. Beach Area 1 is north of the Pier (33° 11′ 54.83″ N, 117° 23′ 16.97″ W). Beach Area 2 is south of the Pier (33° 12′ 7.7″ N, 117° 23′ 29.6″ W). This beach was the site of the sand replenishment project (see Figure 1).

In 1998, a method for assessment of grunion runs, the Walker Scale, was developed that evaluates the number of fish on shore, the extent of shoreline involved, and the duration of the run [30], shown in Table 1. This has been in use in conjunction with a group of trained citizen scientists, the Grunion Greeters, since 2002 [31]. Their reports are reliable and consistent [25]. Since they are volunteers, the beaches they observe and the number of nights they go out are not as regular as one might wish. Nevertheless, Oceanside City and Oceanside Harbor Beaches have both been monitored for spawning runs on many occasions by Grunion Greeters since 2004.

Table 1. The Walker Scale scoring system. Grunion Greeters watch a given stretch of beach for 2 h beginning at the highest nightly tide, usually within the first four days following a new or full moon.

W0:	No fish or only a few individuals appear, with little or no spawning; not a run
W1:	Between 10–100 fish present on the beach over the time of the run, in one or more locations, with little spawning; poor run
W2:	During the peak of the run, 100–500 fish on the shore simultaneously, spawning in one or more locations along the beach; small run
W3:	During the peak of the run, hundreds to thousands of fish spawning at the same time in one or several areas of the beach; peak is less than 20 min; good run
W4:	During the peak of the run, thousands of fish are on the beach together, with little sand visible between them, in a restricted or large area of the beach. Peak lasts less than one hour.
W5:	At the peak of the run, fish fully cover an extensive area of the beach in massive numbers, several individuals deep, a silver lining along the surf. It is not possible to walk through the run without stepping on a fish. The peak lasts over an hour.

Using the Walker Scale, a section of beach is observed during a night when a run may occur (www.Grunion.org) [22], four nights following a new or full moon. In many cases, with beaches that extend a long distance along the shore, only part of the beach may be involved in a run on a given night [15,16]. Once a run starts, it may continue for several minutes to over an hour, however, the peak of the run generally lasts between 20 min and 1 h. Runs typically do not continue beyond two hours after the tide, as the ebb makes the proper tidal placement of the eggs difficult [15].

In 2016, 2017 and 2018, Oceanside Harbor and City Beaches were observed by CDFW staff and Grunion Greeters on nights when the grunion were predicted to spawn. Professional biological monitoring staff were contracted as required for the USACE permits to observe for grunion spawning during the project activities. Their reports were compared with our reports for the same dates and provided data for additional dates. Archival Grunion Greeter data were accessed for past grunion run reports.

The grunion run on the sloped beach face in the intertidal zone, riding waves into the upper intertidal zone. Scarps are sharp steep drops in the beach face that can block access to the upper intertidal zone, where grunion normally spawn. Formation of scarps was documented with photos and by measuring height and length. Beaches were compared for both the changes to the beach face and grunion spawning activity. Photos were taken of spawning on cobble and sandy beaches, as well as next to the discharge pipe and other activities related to beach filling.

3. Results

3.1. *Effects on Sand Habitat*

Beach-fill projects took place on Oceanside City Beach from 6 June to 31 October. The dredge pipe length on the Oceanside Harbor beach was buried parallel to the shore in the dry beach sand above the higher high tide. A ten-inch diameter pipe extended south across the San Luis Rey River and ran on top of Oceanside City Beach. Approximately 50,000 cubic yards were deposited at the first discharge point in Area 1. About 1000 feet farther south, 10,000 cubic yards were added. Finally, 190,000 cubic yards were added just north and south of the Oceanside Pier (Figure 1), for a total of 250,000 cubic yards of material.

The northern half of Area 1 on Oceanside City Beach initially showed very little upper intertidal habitat suitable for grunion spawning due to cobble mounds. Adequate upper intertidal habitat was initially present on Area 2, from Oceanside Pier South 1300 feet down the coast. In 2016, no new scarps developed after beach filling (see photo Figure 3a).

(a) (b)

Figure 3. (**a**), Before beach filling, 28 March, 2017, Area 2 near the Oceanside Pier had a low scarp. Right, (**b**), Night of 26 June, 2017 in Area 1 showed a W3 spawning intensity, the highest seen on Area 1 that season. Beach filling blocked habitat in Area 2, diverting fish to Area 1. Photos by L. C. Adams.

In 2017, before beach fill, the northern half of Area 1 and the southern-most portion of Area 2 were mostly cobble or gravel. Closer to the pier, adequate grunion spawning habitat with flat, sandy beach existed prior to beach fill (Figure 3b).

In 2017 beach fill took place again during the peak of grunion season, from 17 April to 11 June. A 28-inch discharge pipe diameter was used to deposit 420,000 cubic yards of sediment. During the project, Area 1 beach developed steep scarps about 4 feet high on the north side adjacent to the pier area, and in the north part of Area 1 (Figure 4). This steep scarp blocked the transition to the upper intertidal zone, where grunion spawning could have otherwise occurred. The beach had a lot of cobble below the scarp, with no sand patches between cobble swales.

(**a**) (**b**)

Figure 4. (**a**), On 12 June 2017 in Area 2 south of Pier, a steep 5-foot scarp formed in upper intertidal. No grunion spawned in this area after the scarp formed. (**b**), After beach fill, on 25 July 2017 in Area 1, the scarp is lined with cobble in upper intertidal. Cobble was present before and after beach fill. Photos by L. C. Adams.

The beach fill project included bull-dozers building a dike and sand discharge that caused a large pit to form due to scouring on the north side of the pier on 17 May 2017. Sand was also deposited onto the south side of the pier (Figure 5). Natural sand transport contributed to the building up of the Area 2 beach. The pits were later filled in. The Area 1 scarp leveled out, but Area 2 spawning habitat was impacted when a new steep 5-foot scarp formed (Figure 6). In 2017, a lot of cobble was still visible on the beach even after completion of the beach fill project.

(**a**) (**b**)

Figure 5. (**a**), On 17 May 2017: Dykes were built up in the foreground, and pits were scoured out near the Oceanside Pier in Areas 1 and 2 during peak grunion spawning season, causing temporary impacts on grunion spawning habitat. (**b**), Area 2 south of Oceanside Pier. Photos by L. C. Adams.

(a)

(b)

Figure 6. (**a**), Area 2 south of Oceanside Pier in March, 2018, showing scarp lingering from the 2017 project. (**b**), Area 1 north of the Pier showing scarps before the 2018 beach fill in October. Photos by L. C. Adams.

In 2018, although beach fill did not occur during the grunion spawning season, tall scarps still lingered in the high intertidal grunion spawning zone from the 2017 beach fill project (Figure 6). In 2018, the beach fill project occurred from 18 October through to 6 November, after grunion spawning season ended. A 28-inch pipe diameter was used for deposition of 285,000 cubic yards of dredged sediments onto Oceanside City Beach. Along with the lingering scarps from 2017, new scarps developed, shown in photos from September 2018. These were contoured flat by bulldozers by the end of the project in November 2018.

3.2. Grunion Spawning Observations

From 2004–2018, Grunion Greeters provided 90 observations of grunion runs on Oceanside City Beach and 82 observations on Oceanside Harbor Beach.

Observers from the contracted project provided Walker scores and observations on 25 nights in 2016 and 13 nights in 2017 on Oceanside City Beach, during nights when runs were forecast, and the project was ongoing. Because the 2018 project took place after the grunion spawning season, observations of grunion were made only by volunteer California Department of Fish and Wildlife staff and Grunion Greeters, not by contracted project staff. Volunteers contributed reports for eight nights in 2016, six nights in 2017, and seven nights in 2018 on the three areas of Oceanside beaches.

The Walker Scale is categorical, not parametric, so scores were compared with non-parametric statistics. Reports from the observations by professional biologists agreed with the reports from the CDFW and Grunion Greeters when both were present in the same areas on the same nights (Mann-Whitney comparison of two groups, range W0–W3, $N = 15$, $U = 112.5$, $z = 0.02$, $p = 0.985$). During the recreational fishing open season, it was difficult to tell how many fish were present in some locations because people caught them as soon as they appeared; often, nearly every fish that approached the shore was taken before it had the opportunity to spawn.

In 2016, runs ranging from W0 to W2 were reported in Oceanside Harbor, with a median of W1. At Oceanside City Beach within the project footprint, the range reported was W0–W1, with a median of W1.

In May 2017, during beach fill, on Area 1 groups of about 5–6 grunion (W0) appeared along the beach in a few locations, near the pipe while discharging a sand slurry directly seaward into the intertidal zone (Figure 3B). In late May, no grunion were sighted on any of the run nights in either project Areas 1 or 2. In July, no grunion were seen on Oceanside Harbor Beach. A spawning intensity of W3 was seen in Area 1 of Oceanside City Beach in July on sand areas in between cobble. No grunion spawning occurred in locations with high scarps. In Area 2, grunion were seen in the water only, with no spawning as a scarp was still present over the intertidal zone. The upper intertidal scarps on the affected beach areas blocked grunion spawning habitat. Waves ran up and hit the face of the scarps, and sometimes reflected back or over-topped the scarp.

All grunion runs in 2018 were scored W2 or lower on both Oceanside Harbor and Oceanside City beaches.

Spawning run scores on Oceanside Harbor Beach were not significantly different from scores on Oceanside City Beach, either in 2006–2008 or in 2016–2018 (Wilcoxon Signed Rank test of paired data for two groups, $W = -13$, $z = -0.44$, $p = 0.66$; Mann–Whitney test of all data, $U = 102$, $z = 0.02$, $p = 0.98$). The majority of runs on Oceanside Harbor and Oceanside City Beaches from 2016–2018 were small, with 86% of nights scored as Walker 0 or 1, and fewer than 100 fish seen during the run, with little or no spawning. No runs at any of these beaches scored as Walker 4 or 5 over these three years.

In 2006–2008, grunion runs on Oceanside beaches were not different from southern California beaches as a whole, but in 2016–2018, both the southern California grunion population and the Oceanside grunion population show declines in run strength. In 2006–2008, the distribution of frequencies of scores of Oceanside's runs were not significantly different from those across southern California ($X^2 = 8.57$, df = 5, $p = 0.128$).

However, comparing reports from Oceanside City and Harbor Beaches in 2016–2018 with those from 10 years earlier, 2006–2008, a significant difference in frequencies of Walker scores was found (Chi-Square, $X^2 = 32.14$, df = 4, $p < 0.0001$, Figure 7). Low Walker scores were reported with significantly higher frequency in both Oceanside beaches in recent years than in the past ($X^2 = 32.14$, df = 4, $p < 0.0001$), with the median dropping from W2 in 2006–2008 to W1 in 2016–2018. Some nights in April, May and June when spawning was forecast, no grunion were observed on affected beaches. The likelihood of seeing few or no fish (W0) in Oceanside on a night when a run is forecast almost quadrupled, from 14% in 2006–2008 to 54% in 2016–2018.

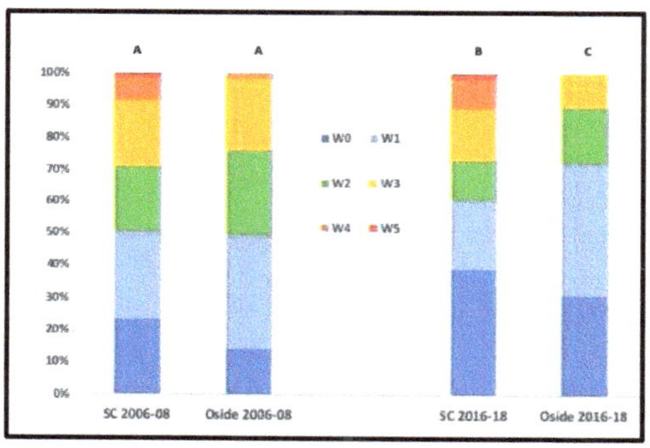

Figure 7. Comparison of grunion run frequencies between the years 2006–2008 and 2016–2018, for reports from all of Southern California (SC) as compared with combined data from Oceanside Harbor and Oceanside City Beaches (Oside). Different letters above the bars indicate significant differences in run frequencies from SC 2006–2008. Small runs were significantly more frequent in 2016–2018 than in 2006–2008.

No runs above W3 have been reported on Oceanside beaches in recent years. By comparison, between 2016–2018 across southern California, large runs (W4 or 5) occurred for 10% of the reports, on many different beaches. Oceanside City and Harbor Beaches in 2016–2018 were significantly lower in frequencies of medium (W2–3) and larger (W4–5) runs as compared with the southern California grunion habitat range as a whole ($X^2 = 77.4$, df = 5, $p < 0.0001$, Figure 7).

4. Discussion

Multiple stressors impact marine organisms, especially those along the coast [21,32,33]. California grunion have never been an abundant species [14]. Currently, they face a dwindling spawning habitat within a limited habitat range, along one of the most densely populated coastal zones of the world [15,18]. Human recreational activities, capture during spawning, and coastal development activities can interfere with the critical portion of their life cycle during reproduction by scaring away spawning fish and interrupting or stopping runs that depend on tidal timing [15,25,26]. California grunion are important in the marine ecosystem and food web [17,34], although they are not state or federally listed. They are a popular sport fish when they come ashore to spawn [18,25,35]).

Grunion Greeters have been monitoring grunion spawning for nearly two decades, in cooperation with the California Department of Fish and Wildlife (CDFW). The stock size is difficult to assess, but it is a restricted resource [14]. In recent years, grunion have expanded their spawning range north of San Francisco [36], but the main population in southern California shows signs of decrease [25]. The California Fish and Game Commission is currently considering changes in the open season for the recreational fishery, which takes place while grunion are spawning.

Spawning runs in Oceanside at both the Harbor and City beaches were scored significantly lower in the past three years than in three years a decade ago (Figure 7). No large runs occurred at Oceanside during the past three years, although some other southern California beaches saw repeated large runs [25]. The decline in runs was seen at both beaches, even though beach filling took place only on one. This suggests that a nearby beach may have negative impacts from a beach filling project even if it is not directly receiving dredged sand. It is likely that the individual fish that spawn at both nearby beaches are part of the same metapopulation [36,37], thus changes to the reproductive success at one beach may affect the other.

In 2016, dredgers deposited less than half the volume placed in 2017. Sediments were placed slowly onto the beach by a 10-inch diameter discharge pipe, and high wave events caused some project delays. Even with El Nino oceanographic conditions and a high amount of beach erosion during 2016, beach scarps did not develop (Figure 3). Dredged sand remained on the beaches with mild slopes throughout the summer and fall. The good sandy beach and spawning habitat conditions present long after the 2016 beach fill may have occurred because the sediments were placed slowly. This may have allowed build-up of beach height and time for waves to re-distribute and winnow away the sediments naturally, keeping the beach sand stabilized with only temporary, low scarp formations (Figure 3).

In spring 2017, before beach fill, good spawning habitat conditions were present on most of the grunion survey areas north and south of the pier. In 2017, a 28-inch diameter pipe was used, and the beach in front of the discharge pipe built up quickly, likely causing beach scarps to form when large quantities of sediment were placed in a short amount of time (Figure 4). After beach filling ended, the 5-foot scarp that formed south of the pier left no upper intertidal spawning habitat to support grunion spawning in Area 2 along 700 feet of beach (Figure 6). The sandy beach habitat around the pier was in better condition before the 2017 fill project, with no scarps and very little cobble, as compared to afterward (Figure 6). Spawning in Area 2 before beach filling was a W3 (Figure 3B), but after beach filling, no spawning was seen.

Sand scarps can create grunion spawning impacts by blocking access to the appropriate tidal height and by eroding and burying nests during the incubation period [3,27]. During the observations throughout the study, Oceanside Harbor Beach never developed significant scarps and remained a mildly sloped, sandy beach (Figure 2), while Oceanside City Beach within the fill zone developed steep, significant scarps blocking access to the tidal zone required for grunion spawning.

Small scarps form naturally on Oceanside City Beach, but beach fill can significantly contribute to the steepness and elevation of the scarp, as well as its tidal height location. A sand cliff, much like a seawall, can cause wave scouring of the intertidal zone [38,39]. A high beach scarp can linger in that condition for many weeks or months, as one did from 2017–2018, until a bulldozer or high waves knock it down (Figure 6).

In 2017, a large pit developed due to discharge scouring effects near the pier (Figure 5), which was later filled in. This had immediate impacts preventing spawning runs in these locations. The scarp in Area 1 leveled out, however, the steep scarp that formed in Area 2 from the rapid sand build-up remained for the rest of the spawning season of 2017 and blocked the grunion spawning zone. In the future, placement of the pipe on a mostly cobble beach could be a less impactful location for sand discharge. Alternatively, depositing sediments in areas in the nearshore within the depth of closure would allow for gradual onshore beach replenishment via waves and currents [40,41].

Animals and plants on California's sandy beaches face multiple stressors, both human-caused and natural [42–44]. Climate change, sand erosion and sea-level rise exacerbate seasonal sand volume changes [45–47]. Efforts to replenish sands and deposit dredge material must take into account the impact of these projects on local beach biota [6,7,44], particularly as these projects become more common and frequent in the future [48,49]. Complicating the issue, many marine organisms are shifting their habitat ranges as climate changes [50], including the California grunion [36,37]. Movement into a new habitat range has led to changes in ecological and life history characteristics that make predicting the future of the California grunion difficult [19,25].

Grunion hatch timing is not consistently predictable. Embryos hatch according to an environmental cue, being washed out of the sand into waves as the tides rise for the next semilunar high tides [51]. Because of this, grunion embryos may remain unhatched but viable on shore for up to four weeks (two semilunar tidal cycles) or more, if not washed free into seawater [52,53]. Because of California's mixed semidiurnal tide regime, the height of the highest tides, on which the grunion spawn, differ greatly between the new and full moons [16,54], and thus all eggs may not be washed back to sea to hatch within a subsequent tidal cycle. However, some beach fill permits allow activity to re-commence in a known grunion spawning site after 10 days (the earliest time that embryos are likely to be viable for release) rather than avoiding the area until after the highest high tides occur.

The presence of numerous anthropogenic disturbances may create synergistic effects on vulnerable organisms and ecosystems, particularly with high disturbance frequency [9,33,43]. The decline of the grunion spawning runs in Oceanside probably is due to multiple stressors, and is similar to the decline seen across southern California. However, in 2006–2008, Oceanside's runs were not significantly different from those seen in the total southern California population (Figure 7). Now, the runs at Oceanside beaches are significantly lower than runs across southern California as a whole. Particularly troubling is the complete absence of large runs (W4–5) in the past few years. This differs from many other grunion beaches, that continued to have large runs between 2016–2018. The long-standing history of frequent sand replenishment at Oceanside may explain this decline in runs today relative to the past, or it may be affected by more recent changes in procedures or timing.

5. Conclusions/Recommendations:

1. Beaches are ecosystems. Beach replenishment projects done for human purposes of harbor dredging have ecosystem effects that are not fully understood and may be negative rather than positive for the ecology. Projects should take care to avoid or minimize disturbing critical habitats during reproductive seasons. Ideally, beach fill in Oceanside and other California beaches should be completed by the end of March to avoid the peak grunion spawning months of April, May and June.
2. The addition of new substrate should be done gradually. The beach should build up slowly over time, allowing for a more natural beach face with a gradual slope rather than a steep scarp. Then, even if high wave events occur, a steep scarp is less likely to form or will have a lower elevation allowing for waves and tides to break down the scarp naturally.
3. Pipe discharge scouring and bulldozing should be avoided on sandy beaches close to piers. Pier locations are hot spots where grunion have historically spawned in high numbers [16,54].
4. Dredge sediments must continue to ensure a grain size similar to the natural sandy beach baseline conditions, in order to provide an appropriate slope of the beach face and suitable habitat for local biota.
5. Frequent repetition of sand placement and harbor dredging may accumulate impacts by not allowing sufficient time for the ecosystem to recover before additional disturbance occurs. Rather than improving habitat, these repeated projects in Oceanside may actually be degrading the spawning habitat for grunion, both at the project site and neighboring beaches. Project impacts

from 2017 lingered through the grunion season of 2018, even though the 2018 project itself started after grunion spawning season.

6. Alternative discharge methods, attention to slope and elevation designs, smaller sediment volumes, less impactful locations for placement, and greater care in beach fill practices should be implemented to reduce future impacts.

Author Contributions: Both authors shared in the conceptualization, methodology, and formal analysis; data curation, K.L.M.M.; writing—original draft preparation, K.L.M.M.; writing—review and editing, K.L.M.M. and L.C.A.; visualization, L.C.A. All authors have read and agreed to the published version of the manuscript.

Funding: This research was funded by the US Fish & Wildlife Service, "Connecting People with Nature," NOAA—USC Sea Grant College—Urban Oceans Program, and Pepperdine University.

Acknowledgments: We thank volunteer observers R. Valluzi, and M. Studer and the Grunion Greeters. R.D. Martin, B. Ota, K. Ramey, C. Valle, and A. Barilotti provided many helpful comments on the manuscript.

Conflicts of Interest: Author L.C.A. is employed by the California Department of Fish and Wildlife. The conclusions and interpretations of the data may not represent the official viewpoint of the Department or the State of California. The funders had no role in the design of the study; in the collection, analyses, or interpretation of data; in the writing of the manuscript, or in the decision to publish the results.

References

1. Leonard, L.; Dixon, K.; Pilkey, O. A comparison of beach replenishment on the U.S. atlantic, pacific and gulf coasts. *J. Coast. Res.* **1990**, *6*, 127–140.
2. Pilkey, O.H., Jr.; Cooper, J.A.G. *The Last Beach*; Duke University Press: Durham, NC, USA, 2014; p. 256.
3. Lawrenz-Miller, S. Grunion spawning versus beach nourishment: Nursery or burial ground? *Coast. Zone* **1991**, *3*, 2197–2208.
4. Peterson, C.H.; Hickerson, D.H.M.; Johnson, G.G. Short-term consequences of nourishment and bulldozing on the dominant large invertebrates of a sandy beach. *J. Coast. Res.* **2000**, *16*, 368–378.
5. Peterson, C.H.; Bishop, M.J. Assessing the environmental impacts of beach nourishment. *Bioscience* **2005**, *55*, 887–896. [CrossRef]
6. Rosov, B.; Bush, S.; Roberts Briggs, T.; Elko, N. The state of understanding the impacts of beach nourishment activities on infaunal communities. *Shore Beach* **2016**, *84*, 51–55.
7. Herrera, A.; Gomez-Pina, G.; Fages, L.; Casa, A.; Muñoz-Perez, J.J. Environmental impact of beach nourishment: A case study of the rio san pedro beach (SW Spain). *Open Oceanogr. J.* **2010**, *4*, 32–42. [CrossRef]
8. Viola, S.; Dugan, J.E.; Hubbard, D.M.; Schooler, N.K. Burrowing inhibition by fine textured beach fill: Implications for recovery of beach ecosystems. *Estuar. Coast. Shelf Sci.* **2014**, *150*, 142–148. [CrossRef]
9. Peterson, C.H.; Bishop, M.J.; D'Anna, L.M.; Johnson, G.A. Multi-year persistence of beach habitat degradation from nourishment using coarse shelly sediments. *Sci. Total Environ.* **2014**, *487*, 481–492. [CrossRef]
10. Mason, T.; Coates, T.T. Sediment transport processes on mixed beaches: A review for shoreline management. *J. Coast. Res.* **2001**, *2001*, 645–657.
11. Muñoz-Perez, J.J.; Medina, R. Profile changes due to a fortnightly tidal cycle. *Coast. Eng.* **2001**, *2000*, 3052–3075.
12. Peterson, C.H.; Bishop, M.J.; Johnson, G.A.; D'Anna, L.M.; Manning, L. Exploiting beach filling as an unaffordable experiment: Benthic intertidal impacts propagating up to shorebirds. *J. Exp. Mar. Biol. Ecol.* **2006**, *338*, 205–221. [CrossRef]
13. Dugan, J.E.; Hubbard, D.M.; Quigley, B. Beyond beach width: Steps toward identifying and integrating dynamic ecological envelopes with geomorphic features and datums for sandy beach ecosystems. *Geomorphology* **2013**, *199*, 95–105. [CrossRef]
14. King, P.G.; Nelsen, C.; Dugan, J.E.; Hubbard, D.M.; Martin, K.L. Valuing beach ecosystems in an age of retreat. *Shore Beach* **2018**, *86*, 1–15.
15. Gregory, P.A. Grunion. In *California's Living Marine Resources: A Status Report*; Leet, W.S., Dewees, C.M., Klingbeill, R., Larson, E.J., Eds.; California Department of Fish and Game: Sacramento, CA, USA, 2001; pp. 246–247.

16. Martin, K.L.M. *Beach-Spawning Fishes: Reproduction in an Endangered Ecosystem*; Taylor & Francis Group, CRC Press: Oxford, UK, 2015.
17. Walker, B.W. A guide to the grunion. *Calif. Fish Game* **1952**, *38*, 409–420.
18. Sandrozinski, A. California grunion. In *Status of the Fisheries Report, an Update through 2011*; California Department of Fish & Wildlife: Sacramento, CA, USA, 2013.
19. Robbins, E. Essential Fish Habitat in Santa Monica Bay, San Pedro Bay, and San Diego Bay: A Reference Guide for Managers. Master's Thesis, Duke University, Durham, NC, USA, June 2006; p. 129.
20. Martin, K.L.M.; Hieb, K.A.; Roberts, D.A. A southern California icon surfs north: Local ecotype of California Grunion *Leuresthes tenuis* (Atherinopsidae) revealed by multiple approaches during temporary habitat expansion into San Francisco bay. *Copeia* **2013**, *2013*, 729–730. [CrossRef]
21. Available online: Https://visitoceanside.org/?s=grunion (accessed on 28 October 2019).
22. Defeo, O.; McLachlan, A.; Schoeman, D.S.; Schlacher, T.A.; Dugan, J.; Jones, A.; Lastra, M.; Scapini, F. Threats to sandy beach ecosystems: A review. *Estuar. Coast. Shelf Sci.* **2009**, *81*, 1–12. [CrossRef]
23. Martin, K.L.M.; Moravek, C.L.; Martin, A.D.; Martin, R.D. Community based monitoring improves management of essential fish habitat for beach-spawning California Grunion. In *Sandy Beaches and Coastal Zone Management: Proceedings of the Fifth International Symposium on Sandy Beaches, Rabat, Morocco, 19th–23rd October 2009*; Travaux de l' Institut Scientifique: Rabat, Morocco, 2011; pp. 65–72.
24. Rich, C.; Longcore, T. *Ecological Consequences of Artificial Night Lighting*; Island Press: Washington, DC, USA, 2013.
25. Jones, A.; Knapp, H.; Peacock, A.; Wakamatsu, L.; Holt, B. *Southern California Water Resources II: Predicting Grunion Migration Patterns and Spawning Areas in Responses to Changes in California's Oceans*; Develop Technical Report; Jet Propulsion Laboratory, National Atmospheric and Space Administration: Pasadena, CA, USA, 2018.
26. Martin, K.L.M.; Pierce, E.A.; Quach, V.V.; Studer, M. Population trends of beach-spawning California grunion *Leuresthes tenuis* monitored by citizen scientists. *ICES J. Mar. Sci.* **2019**, *76*. [CrossRef]
27. Byrne, R.; Bernardi, G.; Avise, J. Spatiotemporal genetic structure in a protected marine fish, the California grunion (*Leuresthes tenuis*), and relatedness in the genus *Leuresthes*. *J. Hered.* **2013**, *104*, 521–531. [CrossRef]
28. Kuhn, G.G.; Shepherd, F.P. *Sea Cliffs, Beaches, and Coastal Valleys of San Diego County: Some Amazing Histories and Some Horrifying Implications*; University of California Press: Berkeley, CA, USA, 1984.
29. McLachlan, A.; Brown, A.C. *The Ecology of Sandy Shores*, 2nd ed.; Academic Press: San Diego, CA, USA, 2006.
30. Martin, K.L.M.; Raim, J.G. Avian predators target beach spawning marine fish, California Grunion, *Leuresthes tenuis*. *Bull. South. Calif. Acad. Sci.* **2014**, *113*, 187–199.
31. Martin, K.; Speer-Blank, T.; Pommerening, R.; Flannery, J.; Carpenter, K. Does beach grooming harm grunion eggs? *Shore Beach* **2006**, *74*, 17–22.
32. Martin, K.; Staines, A.; Studer, M.; Stivers, C.; Moravek, C.; Johnson, P.; Flannery, J. Grunion Greeters in California: Beach-spawning fish, coastal stewardship, beach management and ecotourism. In *Proceedings of the 5th International Coastal & Marine Tourism Congress: Balancing Marine Tourism, Development and Sustainability*; Lück, M., Gräupl, A., Auyong, J., Miller, M.L., Orams, M.B., Eds.; New Zealand Tourism Research Institute: Auckland, New Zealand, 2007; pp. 73–86.
33. Schoeman, D.S.; Schlacher, T.A.; Defeo, O. Climate-change impacts on sandy-beach biota: Crossing a line in the sand. *Glob. Chang Biol.* **2014**, *20*, 2383–2392. [CrossRef]
34. Rogers-Bennett, L.; Catton, C.A. Marine heat wave and multiple stressors tip bull kelp forest to sea urchin barrens. *Sci. Rep.* **2019**, *9*, 15050. [CrossRef]
35. Spratt, J.D. The amazing grunion. In *Marine Resource Leaflet No. 3*; California Department of Fish and Game: Sacramento, CA, USA, 1986.
36. Roberts, D.; Lea, R.N.; Martin, K.L.M. First record of the occurrence of the California Grunion, *Leuresthes tenuis*, in Tomales Bay, California; a northern extension of the species. *Calif. Fish Game* **2007**, *93*, 107–110.
37. Johnson, P.B.; Martin, K.L.; Vandergon, T.L.; Honeycutt, R.L.; Burton, R.S.; Fry, A. Microsatellite and mitochondrial genetic comparisons between northern and southern populations of California Grunion *Leuresthes tenuis*. *Copeia* **2009**, *2009*, 467–476. [CrossRef]
38. *San Diego Coastal Regional Sediment Management Plan*; San Diego Area Governments (SANDAG): San Diego, CA, USA, 2009; Volume 4, p. 60.

39. Griggs, G.B.; Patsch, K.B.; Savoy, L.E. *Living with the Changing California Coast*; University of California Press: Berkeley, CA, USA, 2005.
40. Thompson, W.F. The spawning of the grunion (*Leuresthes tenuis*). *Calif. Fish Game* **1919**, *5*, 1–27.
41. *California Department of Boating and Waterways and State Coastal Conservancy*; California Beach Restoration Study: Sacramento, CA, USA, 2002.
42. Dugan, J.E.; Hubbard, D.M.; Rodil, I.F.; Revell, D.L.; Schroeter, S. Ecological effects of coastal armoring on sandy beaches. *Mar. Ecol.* **2008**, *29*, 160–170. [CrossRef]
43. Revell, D.L.; Dugan, J.E.; Hubbard, D.M. Physical and ecological responses of sandy beaches to the 1997-98 El Nino. *J. Coast. Res.* **2011**, *27*, 718–730.
44. Orme, A.R.; Griggs, G.B.; Revell, D.L.; Zoulas, J.G.; Grandy, C.C.; Koo, H. Beach changes along the southern California coast during the 20th century: A comparison of natural and human forcing factors. *Shore Beach* **2011**, *79*, 38–50.
45. Schooler, N.K.; Dugan, J.E.; Hubbard, D.M.; Straughan, D. Local scale processes drive long-term change in biodiversity of sandy beach ecosystems. *Ecol. Evol.* **2017**, *7*, 4822–4834. [CrossRef]
46. Yates, M.L.; Guza, R.T.; O'Reilly, W.C.; Seymour, R.J. Overview of seasonal sand level changes on southern California beaches. *Shore Beach* **2009**, *77*, 39–46.
47. Zhang, K.; Douglas, B.C.; Leatherman, S.P. Global warming and coastal erosion. *Clim. Chang.* **2004**, *64*, 41–58. [CrossRef]
48. Vitousek, S.; Barnard, P.L.; Limber, P.; Erikson, L.; Cole, B. A model integrating longshore and cross-shore processes for predicting long-term shoreline response to climate change. *JGR Earth Surf.* **2017**, *122*, 782–806. [CrossRef]
49. Speybroeck, J.; Bonte, D.; Courtens, W.; Gheskiere, T.; Grootaert, P.; Maelfait, J.-P.; Mathys, M.; Provoost, S.; Sabbe, K.; Stienen, E.W.M.; et al. Beach nourishment: An ecologically sound coastal defence alternative? A review. *Aquat. Conserv. Mar. Freshw. Ecosyst.* **2006**, *16*, 419–435. [CrossRef]
50. Flick, R.E.; Ewing, L.C. Sand volume needs of southern California beaches as a function of future sea-level rise rates. *Shore Beach* **2009**, *7*, 36–45.
51. Miller, E.F.; McGowan, J.A. Faunal shift in southern California's coastal fishes: A new assemblage and trophic structure takes hold. *Estuar. Coast. Shelf Sci.* **2013**, *127*, 29–36. [CrossRef]
52. Griem, J.N.; Martin, K.L.M. Wave action: The environmental trigger for hatching in the California grunion, *Leuresthes tenuis* (Teleostei: Atherinopsidae). *Mar. Biol.* **2000**, *137*, 177–181. [CrossRef]
53. Smyder, E.A.; Martin, K.L.M. Temperature effects on egg survival and hatching during the extended incubation period of California grunion, *Leuresthes tenuis*. *Copeia* **2002**, *2002*, 313–320. [CrossRef]
54. Moravek, C.L.; Martin, K.L. Life goes on: Delayed hatching, extended incubation, and heterokairy in development of embryonic California grunion, *Leuresthes tenuis*. *Copeia* **2011**, *2011*, 308–314. [CrossRef]

Journal of
Marine Science and Engineering

MDPI

Article

An Integrated Coastal Sediment Management Plan: The Example of the Tuscany Region (Italy)

Enzo Pranzini [1], Irene Cinelli [1], Luigi E. Cipriani [2] and Giorgio Anfuso [3,*]

1 Department of Earth Science, University of Florence, 50121 Firenze, Italy; enzo.pranzini@unifi.it (E.P.); irene.cinelli@hotmail.it (I.C.)
2 Direction of Soil Defense and Civil Protection, Region of Tuscany, Region of Tuscany, 50127 Firenze, Italy; luigi.cipriani@regione.toscana.it
3 Department of Earth Science, Faculty of Marine and Environmental Sciences, University of Cadiz, CASEM, 11510 Puerto Real, Cádiz, Spain
* Correspondence: giorgio.anfuso@uca.es

Received: 29 November 2019; Accepted: 5 January 2020; Published: 10 January 2020

check for updates

Abstract: This paper presents the results of a study carried out to support the Region of Tuscany Coastal Sediment Management Plan, with the main aim of establishing the sediment budget considering the time span from 1981–1985 to 2005 for the 56 coastal sectors into which the 215 km-long continental sandy coast of Tuscany (Italy) was divided. The sand stability (according to a stability index) and colour compatibility (according to the CIEL*a*b* colour space with an acceptability range conforming to national guidelines) were determined in order to assess the possibility of using the available sediment in accreting sectors to nourish the beach in eroding areas. Only in two cases—i.e., the updrift of a harbour (at Viareggio) and in a convergence zone (at Marina di Pietrasanta)—are the volumes of sufficient magnitude to support a large nourishment project; however, the mean sand size is too small to guarantee efficient nourishment, even with medium-term stability. In contrast, the colour difference, in most of the cases, was shown to be acceptable. Other small sediment stocks, suitable for colour but not for grain size, can be used for periodic ephemeral nourishment works to support seasonal tourist activities. The limited resources available make it necessary to adopt a plan for their optimal use from a regional perspective. This kind of study is of great interest for the proposal of sound management actions to counteract the increasing erosion processes linked to climate change phenomena and human effects on rivers and coastal systems.

Keywords: key coastal erosion; sediment budget; fill stability; colour compatibility; beach nourishment

---

1. Introduction

Tourism is one of the most important industries in the world; global international tourist arrivals grew by 3.9% (1235 million people) in 2016 and 7% (1326 million people) in 2017 [1,2]. Beaches make up a major part of this market [3,4], especially along the Mediterranean coast, which is characterized by mild temperatures associated with annual precipitation in winter and a hot, dry season in summer, which is very attractive for "the Sun, Sea and Sand (3S) tourism" market [5].

Beach erosion, which is a relevant threat to 3S tourism [6], is the result of a deficit in the coastal sediment budget due to the prevalence of outputs over inputs. This process can be locally countered by reducing debits—i.e., sand loss due to erosion processes—by putting shore protection structures in place; however, these increase the deficit in downdrift coastal sectors [7]. Input credits can be increased, favouring soil erosion and river transport or soft rock cliff erosion; both strategies are poorly-suited to developed areas, such as the Tuscan coast in north-western Italy.

The direct injection of sediment quarried on land or extracted from the sea floor to the coast, by means of artificial nourishment works, is the most used approach to increase sediment input [6,8,9].

However, coastal segments where the sediment budget is positive exist; they are observed in both natural conditions, e.g., when sediments are accumulated near headlands at the end of coastal cells or in longshore transport convergence zones [10], and in human-created conditions, i.e., updrift, or in correspondence with coastal structures such as harbours, jetties and breakwaters [11,12].

The sediment circulation framework and sediment budget at the regional scale have to be characterized [13–16] based on the sediment volume continuity equation, which can be calculated both in an analytical way, e.g., with empirically-derived equations [13,17–19] by propagating waves in an area with well-known coastal morphologic and sedimentological characteristics, or by measuring eroded and accumulated sediment volumes, with the latter being mostly relevant to coastal structures [20] or at converging cells limits [13,21]. The former method considers "sediment potential transport", and is easily applicable when nearshore morphological, sedimentological and wave climate data are available [22], whereas the latter deals with actual sediments, but needs accurate historical topographic surveys of the emerged and submerged beach [12,23].

The importance of coastal sediments as a resource for the sound management of the coastal environment and for countering erosion and the effects of climate change has increasingly taken on a strategic value in recent years [24]. This is especially true for Mediterranean coastal areas, which are strongly suited to tourism, and are often characterized by excellent environmental and cultural value, which are strong drivers of economic growth. In this scenario, the management of sediment resources entails the need for specific regional policies for the protection and sustainable use of the different sources and of strategic reservoirs, as underlined by the "EUROSION" project recommendations [25], and by the indications of the "National Guidelines for Coastal Erosion in Italy" [26]. The different natures of the diverse sources involved result in the need to identify adequate management, regulatory and authorization systems in order to favour the sustainable use of sediment resources in a complementary and synergistic way for the purposes of coastal management and protection, along with the principles of the Protocol on Integrated Coastal Zone Management in the Mediterranean [27] for sustainable growth combined with the objectives of territorial safety and the protection of the coastal environment.

A sound regional sediment budget can address shore protection strategies based on sediment bypass from accreting to eroding sectors, which is being carried out though a Regional Sediment Management (RSM) Project developed to implement adaptive management strategies across multiple projects, optimizing the use of sediment while supporting sustainable solutions to navigation and dredging, flood and storm damage reduction, and environmental enhancement missions [16,28,29]. Particular attention also has to be devoted to the effects on beach stability of the extraction of sediments from the nearshore [30].

The coast of Tuscany has experienced significant erosion rates in past decades (Figure 1). To counteract this trend, the Regional Administration planned to quantify and characterize the sediment availability in the coastal area, consisting mainly of sand accumulated updrift of major ports and marinas, and which could allow for significant bypass interventions to nourish eroding sectors. In parallel, sand stocks on the shoreface were analysed which could be used to carry out seasonal beach nourishments to support seaside tourism activities, which represent a significant part of the regional economy. In order to quantify these resources, characterize sediments and assess their suitability for the different coastal sectors experiencing shoreline retreat, a project founded by the Region of Tuscany was carried out at the University of Florence in 2015.

Several aspects have to be taken into account to properly develop a management plan; e.g., designing an adequate, stable artificial beach profile [31–35] and choosing how to suitably borrow sediment, which has to be compatible with the natural one according to textural [23,36] and chromatic characteristics [37–39]. These two aspects have been analysed in this paper.

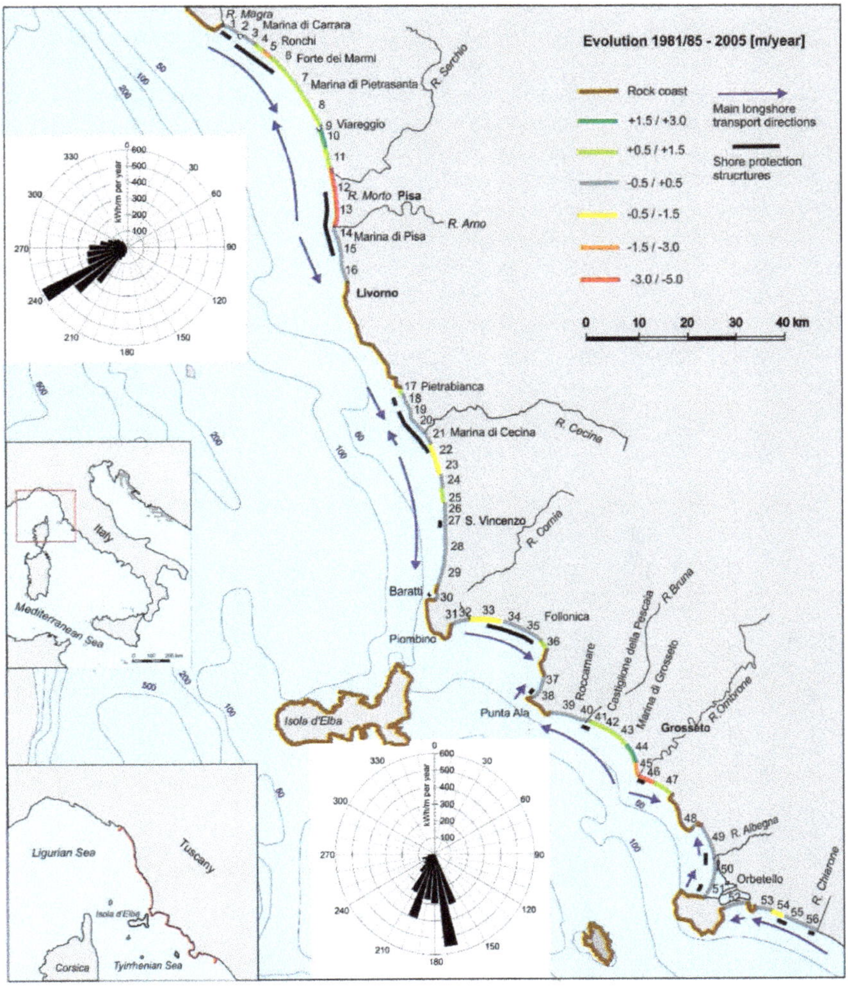

Figure 1. Shoreline evolution along the Tuscan coast from 1981–1985 to 2005; the sectors into which the coast has been divided for the present study are also shown. In Appendix A, the time interval for each coastal sector is presented. Wave energy (KWh/m) directional distribution roses obtained from KNMI data are also reported.

2. Study Area

The continental coast of the Tuscany Region, a microtidal environment, is approximately 380 km long, 215 km of which comprises sandy beaches, with few segments characterized by mixed sand and gravel sediments. This calculation excludes 9 km, or 4.0% of the total coastline, which once consisted of sand beaches but today is protected by seawalls, usually emplaced in correspondence of residential areas and coastal roads, or hosts port facilities [40,41].

The longest continuous offshore wave data series for this area is given by the Koninklijk Nederlands Meteorologisch Instituut (KNMI); wave energy (KWh/m) directional distribution roses derived from 1961–1990 wave data are reported for the areas north and south of Elba Island (Figure 1). In the northern area, wave energy essentially comes from south-west, and in the southern one, from the south.

For more than a century, the Tuscan coast (Figure 1) has experienced an erosive process that, despite the implementation of various counter measures, continues to expand [41]. The coastal evolution in Tuscany was obtained by comparing two shorelines: from1981–1985 and from 2005 [42]. The 1981–1985 shoreline was obtained by means of aerial photographs at an original scale of 1:13,000 with an accuracy of 5 m [40], and the 2005 shoreline was obtained by means of accurate DGPS surveys.

Coastal erosion/accretion rates (m/yr) for the investigated period are shown in Figure 1; the presented values comprise mean shoreline changes in 821 segments, each of which is approximately 250 m long. At each segment, the mean shoreline displacement value (m) was computed by dividing beach area variation by segment length [43] using Geographic Information System (QGIS) tools.

In the period from 1981–1985 to 2005, 9.1% of the regional beaches underwent sever erosion, with retreat rates of more than 3 m/yr; another 12.0% suffered less intense erosion, where beach retreat was between 0.5 and 3.0 m/yr; 27.0% had a shoreline retreat that was slower than 0.5 m/yr (Figure 2). This latter amount, as well as that corresponding to slow (<0.5 m/yr)-accreting areas (+21.8%) were within the mapping accuracy and intrinsic beach variability, and hence have been considered stable (Figure 2).

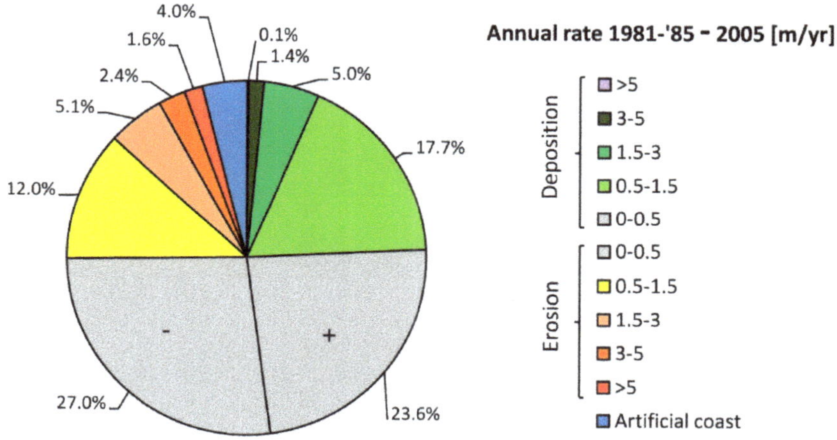

Figure 2. Synthesis of the shoreline displacement rate of the Tuscan coast from 1981–1985 to 2005 based on an unpublished report made by the University of Florence for the Tuscany Region.

The main cause of coastal retreat is the drastic reduction in past centuries of sedimentary inputs by rivers flowing onto this coast and the adjacent regions due to reforestation within watersheds, riverbed quarrying and the construction of weirs and dams [42]. This was reflected by the great retreat observed at river deltas, e.g., at the Arno and Ombrone River mouths (Sectors 12, 13, 45 and 46; see Figure 1). On the northern Tuscan coast, the Magra and Serchio river supplies are approximately 632,000 and 23,000 t/yr, respectively [44,45], and the R. Arno, the most important of the region, contributes approximately 1,524,000 t/yr of the total load [46], which is far lower than that of the previous centuries, i.e., only 37% of that estimated for the 1500–1800 AD period [47].

In the central Tuscan coast, the total load of the Cecina River is approximately 250,000, with half of that being bedload. In the south, the Ombrone River total load is estimated to be 1.35×10^6 m^3/yr [45], far lower than the 5×10^6 m^3/yr estimated by [45,48] for the XVI–XVIII centuries. For the other small water courses emptying onto the Tuscany coast, no bedload data are available, but previous authors have considered all of them (except R. Albegna, which has a drainage area of 749 km^2) to be insignificant to the sediment input.

A minor volume on the regional scale, but important at the local level, was lost in harbour dredging projects to respect the environmental legislation, which led to sediments being deposited offshore or in a Confined Disposal Facility area. In one case, the port authority was asked to compensate the dredging operations by nourishing a downdrift area with sediments from land deposits [49]. Shore protection projects, which have been carried out since the end of the 19th century at river mouths but at other places too (Figure 1), locally reduced the sediment deficit, limiting the debits but also favouring the shift of beach erosion to the downdrift coastal sectors [50,51] according to the "domino effect" [52].

Littoral cells, determined via petrographic analyses [53], explain the impact of these works and also the presence of few depositional areas, the most important being the one observed at Marina di Pietrasanta (Sector 7), where sediments from the Magra River meet with those from the Arno River (Figure 1). The southern littoral cell is fed by sediments delivered by Albegna River, emptying onto the Latium coast and draining a volcanic area; the sand here is the darkest of that of all the other regional beaches.

3. Materials and Methods

Data acquired by the Department of Earth Sciences of the University of Florence during recent decades were used for the computation of available sediment volumes and the assessment of their suitability for beach nourishment. In this paper, firstly, coastal segments with similar morphologies (pocket beaches, barriers, urban vs. rural, shore protection types) and evolution trends (accreting or eroding) were joined into 56 morphologically-uniform sectors ranging in length from 875 to 6946 m, the former corresponding to a beach south of Marina di Cecina delimited by a groin and a creek mouth (Sector 22), and the latter to the Tombolo di Feniglia (Sector 52), the southern tombolo closing Orbetello lagoon (Figure 1).

Secondly, the sedimentary budget was calculated (Appendix A) for each of the sectors (Figure 1). Bathymetric surveys were carried out in 1997–2005 and extended up to a water depth of 10 m along profiles spaced 250 m apart, or 50 m in correspondence of shore protection structures. Depth accuracy, evaluated on several Sea Control Points [54], was 5 cm. For each sector, volume change was calculated considering a prism with dimensions "l" (sector length), "x" (shoreline displacement), "y" (depth of closure), as in Figure 3.

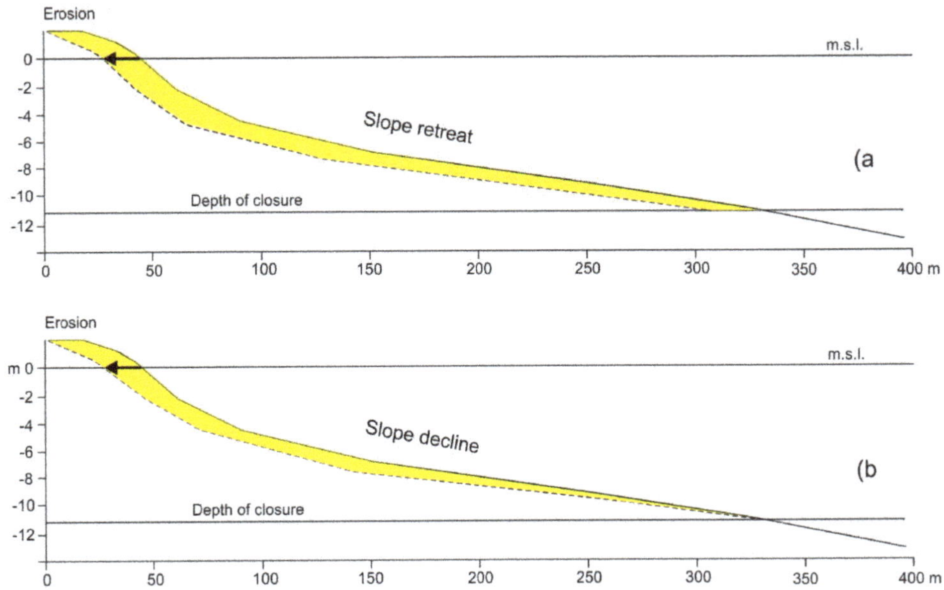

Figure 3. Parallel retreat (**a**) and slope declining (**b**) consequent to a beach sediment deficit.

Finally, instead of a profile parallel retreat model (Figure 3a), a slope decline process was considered, i.e., the erosion and accretion along one profile have an area equal to a triangle whose base corresponds with the shoreline displacement and height equal to the depth of the closure (d) obtained by means of the Hallermaier [55,56] equation (Figure 3b):

$$d = 2.28\ H_s - 68.5\ (H_s^2/gT^2) \tag{1}$$

where H_s is the effective wave height just seaward of the breaker zone that is exceeded for 12 h per year, T is the associated wave period and g is the acceleration of gravity. This choice is supported by the assumption that the upper shoreface responds on a much smaller time scale than the lower shoreface, and the idea that the shoreface profile is not always and everywhere in equilibrium with its forcing [57].

For the coast of Northern Tuscany (65 km long), such a model was also supported by [51], who found steeper profiles in accreting areas and milder ones in eroding areas, as dictated by the inability of the depth of closure point to migrate. In addition, a comparison between bathymetric maps in front of the three most important Tuscany river mouths from 1881 and 1976 showed nonsignificant depth changes (i.e., within the accuracy of the method) offshore of −10 m contour depth [47], which is the depth of closure for a return period of a 50 yr storm [55,58]. Obviously, the volumes (positive and negative) are double if a parallel profile retreat model is considered. For the sediments present in the sectors characterized by a well-established sediment surplus, a hypothesis of their use for the nourishment of eroding sectors was elaborated. For these cases, the suitability of sediments in the deposition areas to fill eroding sectors, both in the dry beaches and in the shorefaces, was evaluated for stability and colour compatibility.

Sediment stability was assessed by computing the Stability index (Si) [23], comparing potential borrowed sediments with the native ones of the eroding beaches. The Si is based on the assumption that the stability of each new grain deposited on the beach is inversely proportional to the percentage of its size in the native sediment cumulative distribution. This data also derives from the archive available at the University of Florence of dry beach and shoreface samples collected in several past projects, upgraded with samples collected for this study, all sieved at a $\frac{1}{2}$ phi interval.

For sand colour assessments and compatibility, the CIEL*a*b* quasi-uniform colour space [37] was used, where coordinates are L* (Lightness), a* (Green–Red axes) and b* (Blue–Yellow axes); the perceived distance between two colours is their Euclidean distance in that space. Measurements were performed with a Konica Minolta CR-410, and the acceptability range was that adopted by the Tuscan Region in several nourishment projects, and recently included in the "National Guidelines for Coastal Erosion in Italy" [26], although small variations were allowed, according to the naturalness of each site:

$$-3 < \Delta L^* < +9 \qquad -3 < \Delta a^* < +3 \qquad -1 < \Delta b^* < +3 \qquad \Delta E^*ab < 10 \tag{2}$$

For each of the 56 coastal sectors, a detailed information form was filled out comprising a site description, photos, shoreline evolution, shore protections (if any), sediment budget, previous nourishments (if any), grain size characteristics and colour determination. For specific sites, additional information is given, e.g., core position, dry beach and shoreface sediment texture maps, etc. (Appendix B).

4. Results and Discussion

In order to counteract coastal erosion [44], actions to increase river sediment input are needed; these include the favouring of soil erosion and the demolition of dams and weirs, as well as the abandonment of the construction of river expansion tanks. All these actions increase flood risk in inhabited areas, and therefore, are opposed by most stakeholders and, consequently, are not proposed by politicians. Hence, the only viable strategy to balance the negative sediment budget is to increase the credits by artificial beach nourishment. Due to the shortage of shelf sediment (relict gravel and sand reservoir [55]) to be used for artificial beach nourishment, and environmental limitations for making use of inland quarried materials, the Region of Tuscany is looking with interest at coastal sediments, at least for small projects and seasonal beach profile maintenance works. Hence, dry beach and shoreface sediment deposits along the Tuscan coast have been assessed for their suitability to nourish eroding sectors, considering their stability for the potential beach destination and colour compatibility with local native sand.

Examples of this strategy exist in USA, with sediment bypassing across barrier island inlets [59,60] in Louisiana [61], and in Australia, with the Tweed River Sediment bypassing, which serves to maintain the Gold Coast beaches and related tourist activity [62]. Finally, beach nourishments carried out in Portugal since 1950 have essentially used (>62%) sediment deposited near or inside harbours [63].

Compared with the traditional systems to define Regional Sediments Budgets, the one here developed is based both on the need to maintain the original beach colour (for landscape and environmental reasons [37,38]) and to guarantee the highest possible fill longevity [23] without strongly modifying the beach itself; this is essentially linked to sand size [39].

On the Tuscan coast, assessment of the sediment budget gives an overallannual deficit of 88,452 m^3, differently spread within the sediment cells of the coast, with only the sectors fed by the Albegna River (Sectors 49, 50, 51) showing a very small surplus (0.6 m^3/m/yr, Figure 4). Identifying the processes driving sediment leak and quantifying related volumes is beyond the scope of the present research, but would be of interest of the Regional Administration, and will be the object of future research. Abrasion is generally considered a minor issue [64], but in mixed sand and gravel sediment beaches, i.e., several sites on the Tuscan coast, it is a significant cause of loss [65]. Landward sediment transport by wind, especially where dunes have been lowered, or vegetation has been cut to create promenades, parking areas and houses, is an effective process of beach sediment loss [12]. On managed beaches, litter and *Posidonia oceanica* leaf removal involves the subtraction of large quantities of sand [66]; this activity is intensively carried out along most of the Tuscany coast. In addition, sand deposited inside harbours is dredged and frequently dumped offshore, as it is not suitable for beach nourishment; a volume of 305,000 m^3 of sand and silt was lost in this way at the Marina di Carrara harbour from 1993 to 2008 [49].

In this paper, the volumes to be extracted have been determined at a few locations (Figure 4). In two cases only (Viareggio, Sector 10, and Marina di Pietrasanta, Sector 7, Figure 4), the volumes were sufficient to justify the implementation of a large nourishment project, but in both, the sand is too fine to be stable on beaches that need additional sediment input, as the Si value shows. In other cases, e.g., the updrift of major groins and jetties (e.g., Pietrabianca, Sector 17, and Marina di Grosseto, Sector 43, Figure 4), limited volumes can justify the utilization of only small mobile bypassing system or road transport to address seasonal sand needs on tourist beaches. Some contexts for the possible use of the results of this study in a regional sediment management program are hereafter described.

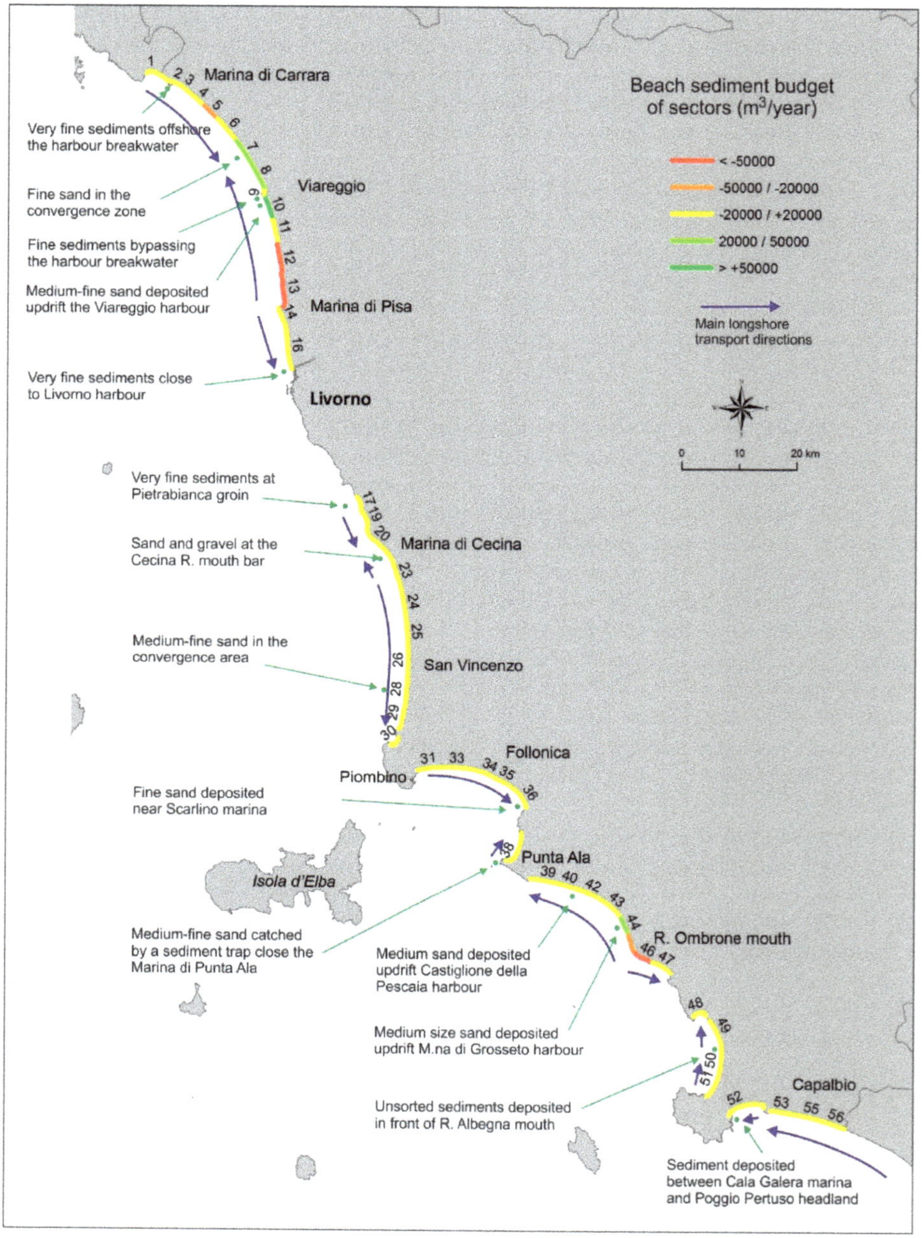

Figure 4. Potential dry beach and shoreface borrow sites for beach nourishment.

4.1. The Beach Updrift of Viareggio Harbour (Sector 10)

The largest sediment deposit of the Tuscan coast is present south (updrift) of Viareggio harbour (Sector 10; Figures 4 and 5), where the progressive expansion of the jetties emplaced (first) at the entrance of the boathouse, and (later) of the harbour, made the beach expand by approximately 630 m from 1878 to 2005. Today, sand partially blocks the harbour entrance, where a bar makes access for boats dangerous; many rescue operations are made each year to assist boats that come to a stop on the shoal and, in 2009, there was a fatal accident.

Figure 5. The updrift breakwater of Viareggio harbour and the beach close to it. Note that the sun umbrella are far from the cabanas, bars and restaurants.

A volume of approximately 1.5 M cubic meters of sand can be excavated to make the shoreline retreat to the 1986 position, leaving a beach large enough to support the intense tourist activity of this site. Actually, the present beach is too wide for optimal tourist use, as sun umbrellas, positioned near the sea, are too far from the cabanas, bars and toilets (Figure 5) [41]. Excessive sand accumulation updrift affecting ports and structures is a common trend in many Mediterranean beaches as observed in Spain [11,12], France [8] and in the south of Italy [67,68].

Due to the thickness of these deposits, 21 two-meter-long cores were taken along the 4.5 km beach updrift of the harbour (Sector 10) to assess sand characteristics of the layer to be used, which proved to have a greater size and colour homogeneity. Unfortunately, the sediments (borrowed sand in Figure 6a,b) were finer than those present in the beaches that needed nourishment work in this sedimentary cell (e.g., Sectors 5 and 12; see Figures 4 and 6). The beach of Sector 5, at Ronchi (native sand in Figure 6a), is more to the north, but it has no fine sediment because it is fed by sand from the north, i.e., the Magra River, that brings coarser sediments. The Si is 0.342, lower than 0.500, which characterizes borrowed sand of equal grain-size distribution to the native one.

Further, the sediment updrift of the harbour (Sector 12, R. Morto beach) are also too fine (native sand in Figure 6b), with a Si of 0.315, to nourish the area from which they originate, i.e., the Arno River mouth (now mostly from the erosion of the river delta) where the shoreline retreats at 3.56 m/yr, with an unitary annual sediment negative budget of 15.3 m^3/m/yr (see Figure 4).

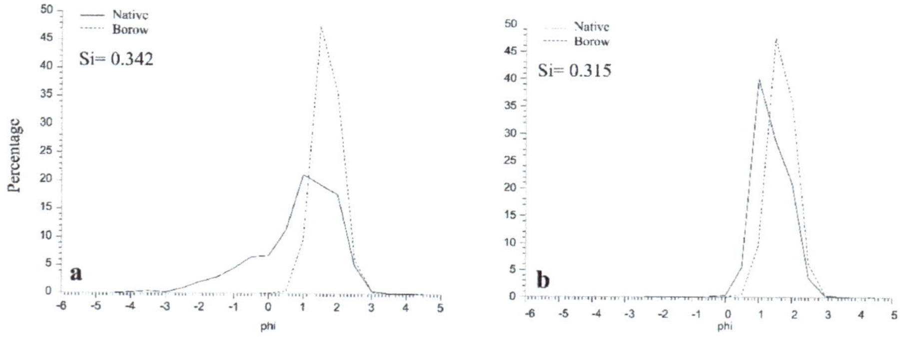

Figure 6. Grain size distribution of the composite sample formed with the 14 cores taken on the 4500-meter-long southern Viareggio Sector (10, borrowed sand) and those of the beach of Sector 5 (**a**, native sand at Ronchi) and 12 (**b**, native sand at R. Morto beach).

Concerning colour characteristics, Viareggio sand (Sector 10) is compatible with that of the areas to be nourished. All the parameters (L*, a*, b*) are within the required limits, and dE*ab is respectively 8.71 and 2.19 (Figure 7). The lower colour difference is with Morto River beach (Sector 12), which is part of the same sedimentary subcell fed by Arno River sediment. The larger differences observed with the Ronchi beach (Sector 5), which poses Viareggio sand near the acceptability range, is given by higher L* and b* values; i.e., the sand is lighter and more yellow, both of which meet with appreciation from the stakeholders. The town of Ronchi has an urban beach, extremely anthropized with bathing establishment and several shore protections structures; and a small change in colour would effect an environmental impact. Changes in sand colour as a result of artificial nourishment have been accepted in different ways by stakeholders; at Varadero (Cuba), where beachgoers are not residents, the darkening of the beach created no problems, whereas on the urban beach of Cagliari (Italy), a similar change of colour led to legal action [38,39]. Although considered unstable, the Tuscany Region has already devised a contract for a first project to dredge 100,000 m^3 of sand from Viareggio to nourish this beach.

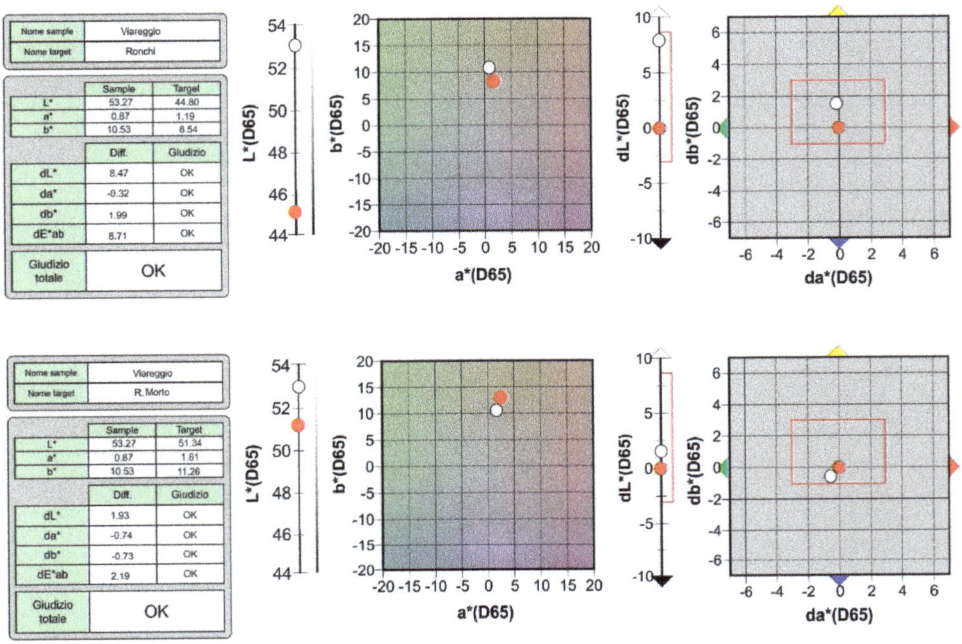

Figure 7. Colour compatibility of Viareggio sand with that of the Ronchi beach (Sector 5) and of R. Morto beach (Sector 12). Acceptability range is delimited in red on dL* axis, with a red rectangle on a*, b* plain. Left graph shows L* and a*b* absolute values of native (withe dot) and borrowed (red dot) sediments; the graphs on the right show the distance of borrowed sediments from native ones.

4.2. Marina di Grosseto (Sector 44)

Another sediment accumulation point along the regional coast, albeit one that is less important than the former, is the Marina di Grosseto recreational harbour (Figure 8), built at the terminal course of an artificial channel in southern Tuscany. The channel, draining the northern Grosseto plain, was excavated in the 1950s and, to prevent the closure of its mouth, a short jetty was built on the southern side. This structure was later lengthened to cope with the growth of the beach, triggering an asymmetry in the shoreline that, in 1998, was 75 m. Originally, the channel was used as a mooring place for small boats but, in 2000, excavation started for a dock and twin curvilinear jetties to prevent entrance silting and allow safe access. Now, the marina can host boats up to 24 m long.

South of the jetty, the beach further expanded by approximately 25 m along a ca. 3156 m segment; the sediment budget for the investigated period is positive, with +7.1 m^3/m/yr, and a mean shoreline displacement of +1.56 m/yr (Figures 1 and 4).

In order to characterize the sediment at Marina di Grosseto, indicated as borrowed sand in Figure 9a,b, 12 cores were taken on the dry beach and on the nearshore bar along a 2 km sector south of the jetty. Their grain size and colour (Figures 9 and 10) were shown to be compatible with those of the two northern, erosive sectors of the sedimentary cell (Sectors 39 and 40) some 10 km away, recording a negative sediment budget of 0.7 and 2.1 m^3/yr respectively (native sand in Figure 9a), and/or the beach to the south near the Ombrone River mouth (Sector 45, native sand in Figure 9b), which is retreating at a rate of 2.31 m/yr, with an annual sediment deficit of 10.5 m^3/m/yr, but with an erosive hotspot of 10 m/yr near the river mouth (Figures 1 and 4).

However, the final destination of this small sand treasure will be decided within a general program of coastal management of the area that the Region of Tuscany is designing on the basis of the present study. The northern sectors are intensively used by tourism, although the coast has an ecologically-sensitive area [69] behind a dune belt. During storms, the dune foot is affected by the run-up and eroded, bringing about the loss of several pines. The beach extension could be useful for environmental and economic reasons although, because the low fill stability (Si = 0.240), nourishment should be considered as ephemeral and would require periodic renourishment works to maintain the beach attractive for tourist activity. If this sand is used to nourish a southern area, close to the Ombrone River mouth, no colour differences will be observed and the fill will result much more stable (Si, i.e., 0.525). However, here, the coast is part of the Maremma Regional Park; it is almost uninhabited and the land is a few centimetres above the mean sea level, and locally below the mean sea level. Therefore, a strategic retreat would be the most sustainable solution for this area.

Figure 8. The boathouse at Marina di Grosseto, whose entrance is protected by two jetties. The updrift one intercepts the long shore sediment transport, creating an extensive beach.

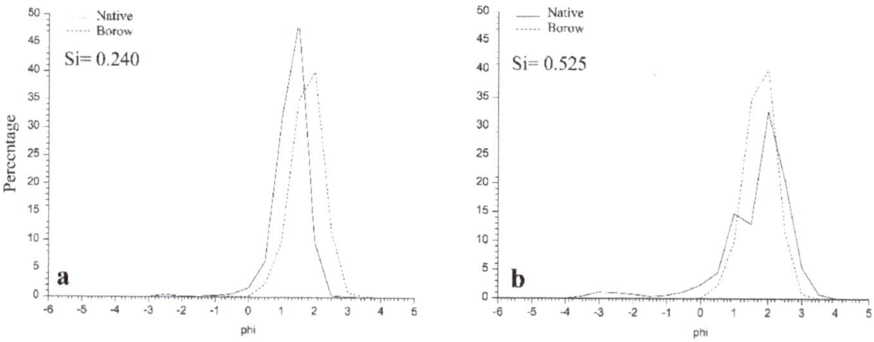

Figure 9. Grain size distribution of the composite sample formed with samples of the borrow area, Marina di Grosseto (12 cores of 4 samples each), compared with those of the potential fill beaches to the northern end of the sedimentary cell (Sectors 39 and 40, **a**, native sand at Roccamare beach) and that representing the beach north of the Ombrone River mouth (Sector 45; 8 samples, **b**).

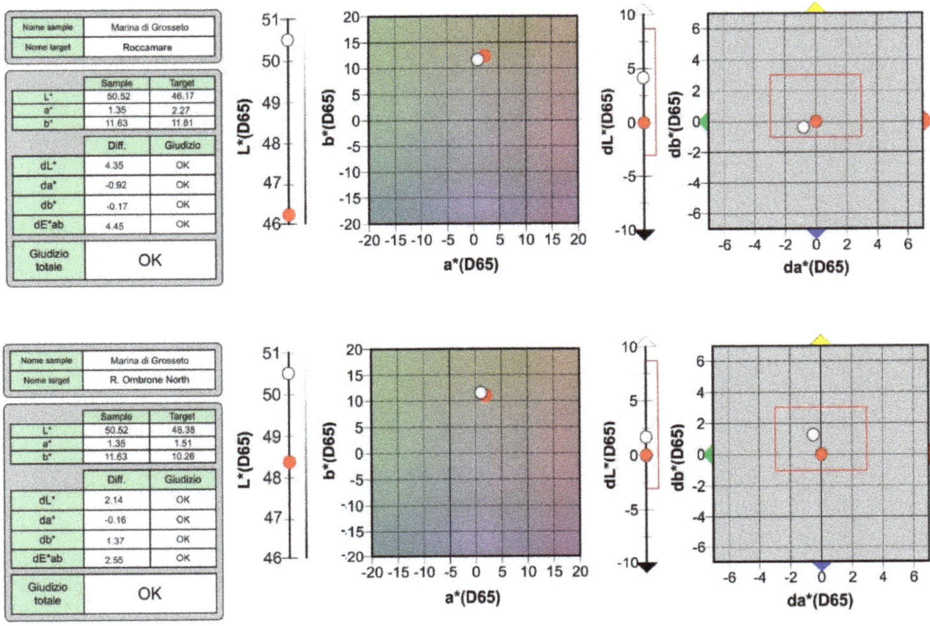

Figure 10. Colour compatibility of Marina di Grosseto sand (Sector 44) with that of Roccamare (Sectors 39, 40) and of Ombrone River North (Sector 45). Acceptability range is delimited in red on dL* axis, with a red rectangle on a*, b* plain. The left graph shows L* and a*b* absolute values of native (with dot) and borrowed (red dot) sediments; the right graphs show distance of borrowed sediments from native ones.

4.3. *Marina di Pietrasanta (Sector 7)*

Larger volumes are available in the convergence area at Marina di Pietrasanta [21], where sediments from the north (brought to the coast by the Magra River) meet those coming from the south (from the Arno River); see Figures 1, 4 and 11. Here, the beach has grown by approximately 250 m since 1878, which is the date of the oldest reliable topographic map; the 1984–2005 sediment budget is + 5.7 m^3/m/yr, with a beach expansion of +1.3 m/yr [23]. Unfortunately, the natural downdrift fining process make this sand the finest and best sorted of all the 65-km-long coastal cell between the Magra River mouth and Livorno [70]; this is a common trend within littoral cells, and has been described by several authors [10,17,21].

The Si of this sand (Figure 12a), if used to nourish the Ronchi beach (Sector 5), is extremely low (Si = 0.065), whereas in the case of the nourishment of the Morto River beach (Figure 12b), the Si is a bit higher (Si = 0.063) but never so much as to influence the evolutionary trend of this beach, not even in the short term.

Figure 11. The beach at Marina di Pietrasanta, in the sediment transport convergence area.

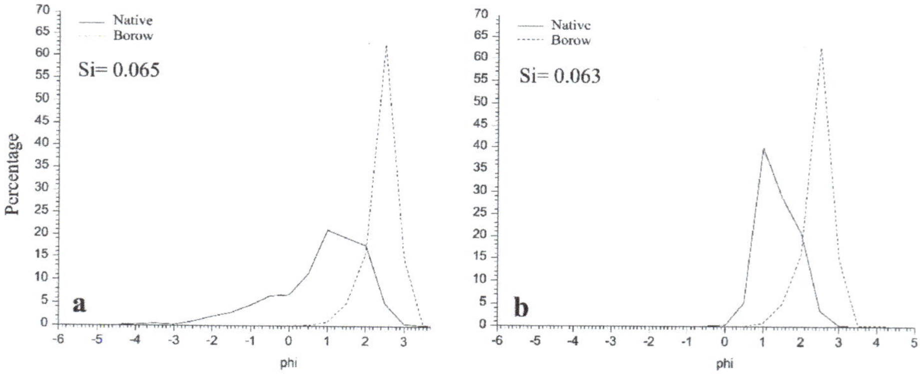

Figure 12. Grain size distribution of the composite sample formed with five samples from Pietrasanta beach (Sector 7, borrowed sand in **a**,**b**) and that representing the native beach sand at Ronchi (Sector 5, **a**) and north of the R. Morto beach (Sector 12, **b**), both averaging three samples.

However, at Marina di Pietrasanta, the entire dry beach is intensively used for tourism activities (Figure 11), and it seems economically and politically impossible to accept a reduction of its surface area. At most, an agreement could be reached with local stakeholders to maintain the present beach width, allowing periodical dredging for the benefit of other coastal sectors, where this sand could help to form/enlarge a beach for the tourist season only.

As far as colour is concerned (Figure 13), compatibility is guaranteed at the Morto River (on the southern subcell), but not at Ronchi (northern sub-cell), suggesting an asymmetry in the longshore transport in the convergence zone.

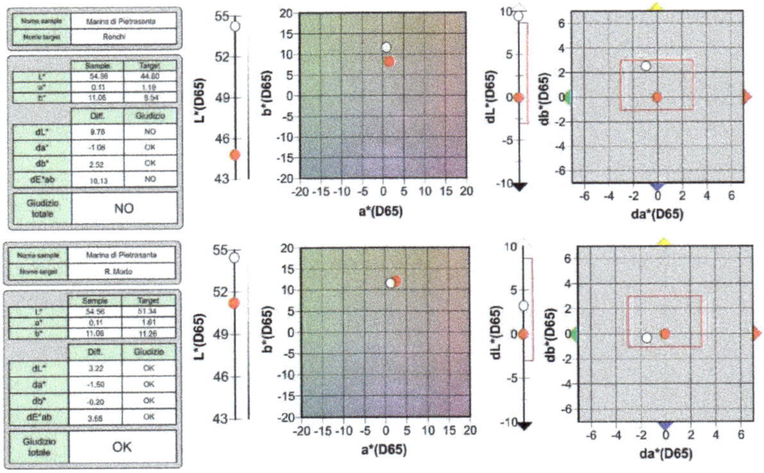

Figure 13. Colour compatibility of Marina di Pietrasanta sand with that of Ronchi beach (Sector 5) and of Morto River beach (Sector 12). Acceptability range is delimited in red on dL* axis, with a red rectangle on a*,b* plain. The left graph shows L* and a*b* absolute values of native (withe dot) and borrowed (red dot) sediments; the right graphs show the distance of borrowed sediments from native ones.

5. Conclusions

Beach monitoring and maintenance is linked to beach function as part of the defence of low-lying coastal areas and assets for the regional tourism industry. On a regional level, approximately 48% of the studied beaches are eroding, notwithstanding the solid protections (seawall, revetments, detached breakwaters, groins) which have been built since the end of the 19th century. Coastal erosion in this region is the result of a sediment deficit trigged by the reduction of river input for land use changes, dam construction, riverbed quarrying, etc., and no actions are planned in the foreseeable future to solve this issue. Furthermore, sea level rise will increase the deficit in coming years.

Sediment artificial input is the only solution to balance the beach budget; this has been performed since the middle of the 20th century using land-quarried aggregates, mostly on the alluvial plains, with the relevant environmental impact. Research to locate, quantify and characterize Holocene shelf deposits has been carried out by the Region of Tuscany, but suitable sediments have not been found yet. Shoreface deposits were therefore considered, assessing their suitability for beach nourishment in eroding sectors joining the same sedimentary cell. Whilst sediment colour poses limited problems, being in many cases compatible with that of the native beaches, grain size is a major issue, since the sands available in depositional areas are finer than those present on eroding coastal sectors.

Sediments which have accumulated updrift of man-made structures or in natural convergent areas can be used to carry out ephemeral nourishment works only, i.e., to create/enlarge narrow beaches of limited temporal durability, which are able to support beach tourist activities for one summer period, but further investigations are needed to define dredging costs, since mob-demob expenses and the distance of the filling area greatly influence sand cost. In any case, the use of sediments does not solve the structural problem of a lack of sand, because the overall sediment budget of the Tuscan coast is negative (−88,452 m^3/yr). Offshore loss due to sediment friction cannot be eliminated, but wiser management of dunes and coastal vegetation could reduce wind landward transport.

A further reduction of sand loss linked to human activity could be achieved by carrying out hand beach cleaning operations and avoiding *Posidonia oceanica* removal. Sediment bypasses at harbour entrances could limit the need for harbour dredging, which negatively impacts beach sediment budget. In any case, a regional coastal sediment management plan, like the one proposed in this paper, constitutes a useful instrument which can assist decision makers and designers in finding cost-effective solutions for shore protection from an integrated coastal zone management perspective.

Author Contributions: E.P. coordinated the research, I.C. made the bibliographic research and wrote the Introduction, L.E.C. made the comparison with similar national and international projects, G.A. reviewed the manuscript and wrote the Conclusions. All authors have read and agreed to the published version of the manuscript.

Funding: This research was founded by the Regione Toscana (Italy).

Acknowledgments: This is a contribution to the PAI Research Group RNM-328 of Andalusia (Spain).

Conflicts of Interest: The authors declare no conflict of interest.

Appendix A

Table A1. Sectors in which the Tuscany continental beaches have been divided and related parameters. Investigated time changes from sector to sector depending on the specific year in which the starting survey was performed, always within the 1981–1985 time span.

Number	Sector Limits	Length (m)	Surface Variation 1981/5–2005 (m^3)	Shoreline Displacement 1981/5–2005 (m)	Time (yrs)	Displacement Rate (m/yr)	Depth of Closure (m)	Annual Sediment Budget (m^3/yr)	Unitary Annual Sediment Budget (m^3/m/yr)
1	Magra–Porto di Carrara	4286	8053.0	1.9	20	0.09	9.1	1832	0.4
2	Porto di Carrara								
3	Porto di Carrara–Marina di Massa	2,250	−13,522.5	−6.0	20	−0.30	9.1	−3076	−1.4
4	Marina di Massa–Foce Frigido	3000	30,471.3	10.2	20	0.51	9.1	6932	2.3
5	Foce Figido–Foce Versilia	3120	−112,816.1	−36.2	20	−1.81	9.1	−25,666	−8.2
6	Foce Versilia–Pineta della Versiliana	5000	47,627.1	9.5	20	0.56	9.1	10,835	2.2
7	Pineta della Versiliana–Lido di Camaiore	5000	126,213.5	25.2	20	1.28	9.1	28,714	5.7
8	Lido di Camaiore–Porto di Viareggio	5000	136,999.3	27.4	20	1.37	9.1	31,167	6.2
9	Porto di Viareggio								
10	Porto di Viareggio–Marina di Torre del Lago	4500	250,576.8	55.7	20	2.78	9.1	57,006	12.7
11	Marina di Torre del Lago–Foce Serchio	4501	59,545.6	13.2	20	0.66	9.1	13,547	3.0
12	Foce Serchio–Foce Fiume Morto Nuovo	5000	−335,888.8	−67.2	20	−3.36	9.1	−76,415	−15.3
13	Foce Fiume Morto Nuovo–Foce Arno	6000	−390,983.7	−65.2	20	−3.26	9.1	−88,949	−14.8
14	Marina di Pisa								
15	Marina di Pisa–Scogliera Milano	1715	−4686.8	−2.7	20	−0.14	9.1	−1066	−0.6
16	Scogliera Milano–Scolmatore	6560	29,781.1	4.5	20	0.23	9.1	6775	1.0
17	Punta Lillatro–Pietrabianca	2115	52,273.4	24.7	24	1.03	9.1	9910	4.7
18	Pietrabianca–Pontile Vittorio Veneto	1080	−7.036.7	−6.5	24	−0.27	9.1	−1334	−1.2
19	Pontile Vittorio Veneto–Pontile Bonaposta	1470	−16,288.4	−11.1	24	−0.33	9.1	−3088	−2.1
20	Pontile Bonaposta–Bocca di Cecina	5419	−4386.7	−0.8	24	0.03	9.1	−832	−0.2
21	Bocca di Cecina–Foce Cecinella	1500	3373.2	2.2	24	0.09	9.1	640	0.4
22	Foce Cecinella–Riserva Tombolo di Cecina	875	−26,651.9	−30.5	24	−1.27	9.1	−5053	−5.8

Table A1. *Cont.*

Number	Sector Limits	Length (m)	Surface Variation 1981/5–2005 (m^3)	Shoreline Displacement 1981/5–2005 (m)	Time (yrs)	Displacement Rate (m/yr)	Depth of Closure (m)	Annual Sediment Budget (m^3/yr)	Unitary Annual Sediment Budget (m^3/m/yr)
23	Riserva Tombolo di Cecina–Marina di Bibbona	5544	−101,043.4	−18.2	24	−0.76	9.1	−19,156	−3.5
24	Marina di Bibbona–Foce Fosso ai Molini	5250	−24,682.5	−4.7	24	−0.20	9.1	−4679	−0.9
25	Foce Fosso ai Molini –La Riconiata	5000	73,514.5	14.7	24	0.61	9.1	13,937	2.8
26	La Riconiata–Porto di S.Vincenzo	4750	7642.7	1.6	24	0.07	9.1	1449	0.3
27	Porto di S.Vincenzo								
28	Porto di S.Vincenzo–Rimigliano	4750	−1893.5	−0.4	24	−0.02	9.1	−359	−0.1
29	Rimigliano–Torre Nuova	4620	753.3	0.2	24	0.01	9.1	143	0.0
30	Golfo di Baratti	2070	−10,700.5	−5.2	21	−0.25	7.8	−1987	−0.96
31	Foce Cornia Vecchia–Torre del Sale	3214	−17,065.0	−5.3	21	−0.25	9.1	−3697	−1.2
32	Torre del Sale								
33	Torre del Sale–Foce Fosso Corniaccia	4500	−47,588.8	−10.6	21	−0.50	9.1	−10,311	−2.3
34	Foce Fosso Corniaccia–Canale Allacciante Corvia	4340	−1250.4	−0.3	21	−0.01	9.1	−271	−0.1
35	Canale Allacciante Corvia–Pineta di Levante	5618	15,304.7	2.7	21	0.13	9.1	3316	0.6
36	Pineta di Levante–Porto di Scarlino	2800	31,801.6	11.4	21	0.54	6.7	5073	1.8
37	Cala Le Donne–Piastrone	2860	4523.8	1.6	22	0.07	7.9	812	0.3
38	Piastrone–Punta Hidalgo	3120	−31,281.3	−10.0	22	−0.46	6.1	−4337	−1.4
39	Punta delle Rocchette–Punta Capezzolo	5750	−56,002.1	−9.7	21	−0.46	9.1	−12,134	−2.1
40	Punta Capezzolo–Foce Bruna	1380	−4578.4	−3.3	21	−0.16	9.1	−992	−0.7
41	Foce Bruna–Limite sud abitato Castiglione	1031	13,336.6	12.9	21	0.62	9.1	2890	2.8
42	Limite sud abitato Castiglione–Pineta del Tombolo	4250	45,594.4	10.7	21	0.51	9.1	9879	2.3
43	Pineta del Tombolo–Porto Marina di Grosseto	4533	75,736.2	16.7	21	0.80	9.1	16,410	3.6
44	Porto Marina di Grosseto–Chiaro del Porciatti	3166	103,897.8	32.8	21	1.56	9.1	22,511	7.1
45	Chiaro del Porciatti–Bocca d'Ombrone	3250	−165.039.7	−50.8	22	−2.31	9.1	−34,133	−10.5

Table A1. *Cont.*

Number	Sector Limits	Length (m)	Surface Variation 1981/5–2005 (m^3)	Shoreline Displacement 1981/5–2005 (m)	Time (yrs)	Displacement Rate (m/yr)	Depth of Closure (m)	Annual Sediment Budget (m^3/yr)	Unitary Annual Sediment Budget (m^3/m/yr)
46	Bocca d'Ombrone–Spiaggia di Alberese	3250	−253,503.3	−78.0	22	−3.55	9.1	−52,429	−16.1
47	Spiaggia di Alberese–Cala Rossa	3750	88,011.7	23.5	22	1.07	9.1	18,202	4.9
48	Golfo di Talamone	2465	−12,796.0	−5.2	21	−0.25	3.8	−1158	−0.5
49	Foce Osa–Bocca d'Albegna	5750	−2767.1	−0.5	21	−0.02	8.3	−547	−0.1
50	Bocca d'Albegna–Tombolo della Giannella	4000	11,972.2	3.0	21	0.14	8.3	2366	0.6
51	Tombolo della Giannella–Santa Liberata	4250	33,376.1	7.9	21	0.37	8.3	6596	1.6
52	Tombolo di Feniglia	6946	56,831.4	8.2	21	0.39	8.7	11,772	1.7
53	Torre Tagliata–Palude di Tagliata	3250	−29,592.3	−9.1	21	−0.43	8.7	−6130	−1.9
54	Palude di Tagliata–Macchiatonda	1500	−33,376.3	−22.3	21	−1.06	8.7	−6914	−4.6
55	Macchiatonda–Palude di Burano	4250	−40,714.4	−9.6	21	−0.46	8.7	−8434	−2.0
56	Palude di Burano–Foce Chiarone	4250	9558.7	2.2	21	0.11	8.7	1980	0.5

Appendix B

Settore 28	Porto di San Vincenzo - Rimigliano		4750 m
Descrizione	Questo settore, lungo 4.750 m, è posto sottoflutto al porto di San Vincenzo e presenta, nel tratto settentrionale, una intensa urbanizzazione, con stabilimenti balneari compressi fra una spiaggia estremamente stretta e il centro abitato. Procedendo verso sud le strutture antropiche si riducono progressivamente, anche se la duna è stata interessata da lottizzazioni e costruzione di villaggi turistici. In molti tratti gli affioramenti di beach-rock sulla battigia e sui fondali antistanti influenzano significativamente l'andamento della linea di riva, in particolare presso La Punticella. Dopo la costruzione del vecchio porto questo settore aveva mostrato un'inversione di tendenza entrando in forte erosione, ma dal 1990 al 2005 il fenomeno si era notevolmente attenuato. Per valutare gli effetti dell'ampliamento del porto sarà necessario attendere qualche anno.		
Trasporto litoraneo	Il trasporto litoraneo netto è diretto verso Sud per un valore prossimo ai 3.000 m^3/anno, con tendenza all'azzeramento verso sud entrando nella zona di convergenza (dati UNIFI-DEAM).		
Strutture costiere	Sottoflutto al porto era stata realizzata una corta struttura sommersa in geocontenitori. Successivamente alla costruzione del porto è stato realizzato un pennello ed una soffolta in scogli che si raccorda ad un altro pennello sommerso con radice emersa. In prossimità de La Punticella è stato realizzato un percorso pedonale con basamento in calcestruzzo sul limite interno della spiaggia in aderenza alla duna.		
Foto	 Spiaggia sottoflutto al porto di San Vincenzo (10 02 2014).	 Percorso pedonale con basamento in calcestruzzo sul limite interno della spiaggia in aderenza alla duna.	

Bilancio 1985-2005	Grafico evol.riva 1938-2005 [m] 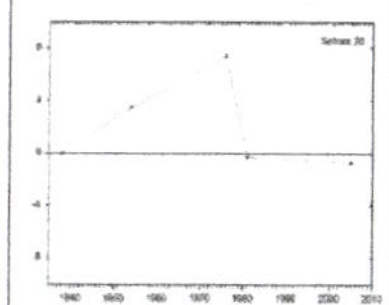	Var. lin. = -0,4 m; Tasso = 0,0 m/anno Var. vol. =-359 m^3/anno Bilancio = -0,08 m^3/m/anno;
Origine dati sedimentologici	UNIFI 2014	
Sedimentologia	Descrizione granulometrica	Mz [phi] / Mz [mm] Spiaggia cella: 1,71 / 0,30 Spiaggia sud: 1,43 / 0,37 Barra: 1,81 / 0,29
Confronto fra la distribuzione granulometrica del campione composito dei sedimenti nativi e quella dei materiali presenti sulla barra a sud del porto.	Ripascimento nella cella 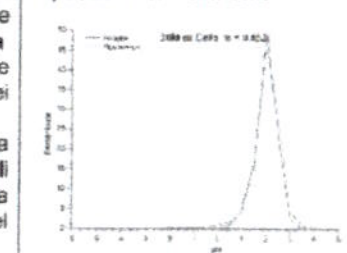	Ripascimento a sud della cella 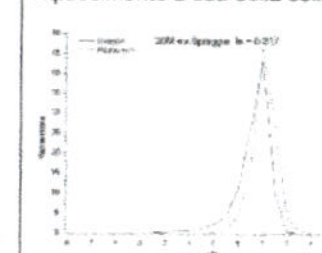
Colore CIEL*a*b*	CIBM 12 LI-B	L* = 53,02 a* = 3,51 b* = 15,02
Utilizzazione o Necessità	Il settore necessita di un consistente intervento di ripascimento per conferire alla spiaggia una forma che consenta l'assorbimento delle mareggiate più intense senza che vengano raggiunte le strutture turistiche. Sedimenti sulla barra sono disponibili, anche a nord del porto, ma di dimensioni non ottimali per un intervento stabile.	

Figure A1. One out of the 56 forms (this is in two pages) reporting information on the sectors in which the Tuscan coast was divided to assess sediment budget, sand characteristics and potentials/needs for artificial nourishment (in Italian, like the original).

References

1. UNWTO. *Tourism Highlights*, 2017th ed.; UNWTO: Madrid, Spain, 2017.
2. UNWTO. *Tourism Highlights*, 2018th ed.; UNWTO: Madrid, Spain, 2018.
3. Klein, Y.L.; Osleeb, J.P.; Viola, M.R. Tourism-generated earnings in the coastal zone: A regional analysis. *J. Coast. Res.* **2004**, *20*, 1080–1088.
4. Houston, J.R. The economic value of beaches e a 2013 update. *Shore Beach* **2013**, *81*, 3–11.
5. Dodds, R.; Kelman, I. How climate change is considered in sustainable tourism policies: A case of the Mediterranean islands of Malta and Mallorca. *Tour. Rev. Int.* **2008**, *12*, 57–70. [CrossRef]
6. Muñoz-Pérez, J.J.; López, B.; Gutiérrez, J.M.; Moreno, L.; Cuena, G. Cost of beach maintenance in the Gulf of Cadiz (SW Spain). *Coast. Eng.* **2001**, *42*, 143–153. [CrossRef]
7. Bruun, P. The Development of Downdrift Erosion. *J. Coast. Res.* **1995**, *11*, 1242–1257.
8. Antony, E.; Sabatier, F. Coastal stabilization practice in France. In *Coastal Erosion and Protection in Europe*; Pranzini, E., Williams, A.T., Eds.; Routledge: London, UK, 2013; p. 457.
9. Hanson, H.; Brampton, A.; Capobianco, C.; Dette, H.H.; Hamm, L.; Laustmp, C.; Lechuga, A.; Spanhoff, R. Beach nourishment projects, practices, and objectives—A European overview. *Coast. Eng.* **2002**, *47*, 81–111. [CrossRef]
10. Bray, M.; Carter, D.; Hooke, J. Littoral cell definition and budgets for central southern England. *J. Coast. Res.* **1995**, *11*, 381–400.
11. Manno, G.; Anfuso, G.; Messina, E.; Williams, A.T.; Suffo, M.; Liguori, V. Decadal evolution of coastline armouring along the Mediterranean Andalusia littoral (South of Spain). *Ocean Coast. Manag.* **2016**, *124*, 84–99. [CrossRef]
12. Molina, R.; Anfuso, G.; Manno, G.; Gracia Prieto, F.J. The Mediterranean Coast of Andalusia (Spain): Medium-Term Evolution and Impacts of Coastal Structures. *Sustainability* **2019**, *11*, 3539. [CrossRef]
13. Bray, M.; Carter, D.; Hooke, J. *Coastal Sediment Transport Study. Report to SCOPAC*; Department of Geography, Portsmouth Polytechnic: Portsmouth, UK, 1991; p. 498.
14. Hooke, J.M.; Bray, M.J.; Carter, D.J. Sediment transport analysis as a component of coastal management—A UK example. *Environ. Geol.* **1996**, *27*, 347–357. [CrossRef]
15. Cooper, N.; Pethick, J. Sediment budget approach to addressing coastal erosion problems in St. Ouen's Bay, Jersey, Channel Island. *J. Coast. Res.* **2005**, *21*, 112–122. [CrossRef]
16. Campbell, T.; Benedet, L.; Fink, C.W. Regional strategies for coastal restoration along the Louisiana Chenier Plain. *J. Coast. Res.* **2005**, *44*, 268–283.
17. Komar, P.D. The budget of littoral sediments, concept and applications. In *Handbook of Coastal and Ocean. Engineering*; Gulf Publishing Co.: Huston, TX, USA, 1990; pp. 681–714.
18. Lowry, P.; Carter, R.W.G. Computer simulation and delimitation of littoral drift cells on the south coast of Co. Wexford, Ireland. *J. Heart Sci. R. Dublin Soc.* **1982**, *4*, 121–132.
19. May, J.P.; Tanner, W.F. The littoral power gradient and shoreline changes. In *Coastal Geomorphology*; Coates, D.R., Ed.; New York State University Press: New York, NY, USA, 1973; pp. 43–61.
20. Hapke, C.H.; Lentz, E.E.; Gayes, P.T.; MCCoy, C.A.; Hehre, R.; Schwab, W.C.; Williams, S.J. A Review of Sediment Budget Imbalances along Fire Island, New York: Can Nearshore Geologic Framework and Patterns of Shoreline Change Explain the Deficit? *J. Coast. Res.* **2010**, *26*, 510–522. [CrossRef]
21. Anfuso, G.; Pranzini, E.; Vitale, G. An integrated approach to coastal erosion problems in northern Tuscany (Italy): Littoral morphological evolution and cells distribution. *Geomorphology* **2011**, *129*, 204–214. [CrossRef]
22. Larson, M. Numerical modelling. In *Encyclopedia of Coastal Science*; Schwarts, M.L., Ed.; Springer: Dordrecht, The Netherland, 2005; pp. 730–733.
23. Pranzini, E.; Anfuso, G.; Muñoz-Pérez, J.J. A probabilistic approach to borrow sediment selection in beach nourishment projects. *Coast. Eng.* **2018**, *139*, 32–35. [CrossRef]
24. Noble, R.M.; Moore, J.T. Coastal Regional Sediment Management Plan. In Proceedings of the Coastal Engineering Proceedings (ICCE), Shanghai, China, 30 June–5 July 2010; McKee Smith, J., Lynett, P., Eds.; Curran Associates, Inc.: New York, NY, USA, 2013.

25. European Communities. *Living with Coastal Erosion in Europe—Sediment and Space for Sustainability—Results from the EUROSION Study*; Doody, P., Ferreira, M., Lombardo, S., Lucius, I., Misdorp, R., NIesing, H., Salman, A., Smallegange, M., Eds.; Office for Official Publications of the European Communities: Brussel, Belgium, 2004; p. 40.
26. MATTM-Regioni. Linee Guida per la Difesa Della Costa Dai Fenomeni di Erosione e Dagli Effetti Dei Cambiamenti Climatici. Versione Settembre 2018—Documento Elaborato Dal Tavolo Nazionale Sull'Erosione Costiera MATTM-Regioni con il Coordinamento Tecnico di ISPRA. Available online: http://www.erosionecostiera.isprambiente.it/linee-guida-nazionali (accessed on 26 December 2019).
27. UNEP/MAP/PAP. Protocol on Integrated Coastal Zone Management in the Mediterranean. Split, Priority Actions Programme. Available online: http://wedocs.unep.org/handle/20.500.11822/1747?show=full (accessed on 26 December 2019).
28. Lillycrop, L.S.; McCornick, J.W.; Parson, L.E.; Chasten, M.A. Adaptive management through regional sediment management. In Proceedings of the WEDA XXXI Technical Conference & TAMU 42 Dredging Seminar, Nashviulle, TN, USA, 5–8 June 2011; pp. 178–187.
29. Finkel, C.W. What might happen to America's shorelines if artificial beach replenishment is curtailed: A prognosis for Southeastern Florida and other sandy regions along regressive coasts. *J. Coast. Res.* **1996**, *12*, 3–9.
30. Kelley, S.W.; Ramsey, J.S.; Byrnes, M.R. Evaluating shoreline response to offshore sand mining for beach nourishment. *J. Coast. Res.* **2004**, *20*, 89–100. [CrossRef]
31. Kana, T.W.; Mohan, R.K. Analysis of nourished profile stability following the fifth Hunting Island (SC) beach nourishment project. *Coast. Eng.* **1998**, *33*, 117–136. [CrossRef]
32. Bodge, K.R. Representing equilibrium beach profiles with an exponential expression. *J. Coast. Res.* **1992**, *8*, 47–55.
33. Capobianco, M.; Hanson, H.; Larson, M.; Steetzel, H.; Stive, M.J.F.; Chatelus, Y.; Aarninkhof, S.; Karambas, T. Nourishment design and evaluation: Applicability of model concepts. *Coast. Eng.* **2002**, *47*, 113–135. [CrossRef]
34. Stive, M.J.F.; de Schipper, M.A.; Luijendijk, A.P.; Aarninkhof, S.; van Gelder-Maas, C.; van Thiel de Vries, J.S.M.; de Vries, S.; Henriquez, M.; Marx, S.; Ranasinghe, R. A New Alternative to Saving Our Beaches from Sea-Level Rise: The Sand Engine. *J. Coast. Res.* **2013**, *29*, 1001–1008. [CrossRef]
35. Bird, E.; Lewis, N. *Beach Renourishment*; Springer: Dordrecht, The Netherlands, 2015; p. 137.
36. Eitner, V. The Effect of Sedimentary Texture on Beach Fill Longevity. *J. Coast. Res.* **1996**, *12*, 447–461.
37. Pranzini, E.; Simonetti, D.; Vitale, G. Sand colour rating and chromatic compatibility of borrow sediments. *J. Coast. Res.* **2011**, *26*, 798–808. [CrossRef]
38. Pranzini, E. Protection studies at two recreational beaches: Poetto and Cala Gonone beaches, Sardinia, Italy. In *Beach Management*; Williams, A., Micallef, A., Eds.; Earthscan Publishers: London, UK, 2009; pp. 287–306.
39. Pranzini, E.; Anfuso, G.; Botero, C.M.; Cabrera, A.; Apin Campos, Y.; Casas Martinez, G.; Williams, A.T. Sand colour at Cuba and its influence on beach nourishment and management. *Ocean Coast. Manag.* **2016**, *126*, 51–60. [CrossRef]
40. Pranzini, E. Beach erosion and protection in Tuscany. In *Monitoring Results: A Capitalization Tool from COASTGAP Project*; Cipriani, L.E., Pranzini, E., Eds.; Nuova Grafica Fiorentina: Firenze, Italy, 2014; p. 138.
41. Pranzini, E.; Anfuso, G.; Cinelli, I.; Piccardi, M.; Vitale, G. Shore Protection Structures Increase and Evolution on the Northern Tuscany Coast (Italy): Influence of Tourism Industry. *Water* **2018**, *10*, 1647. [CrossRef]
42. Bartolini, C.; Cipriani, L.E.; Pranzini, E.; Sargentini, M. Caratteristiche geomorfologiche ed evoluzione della linea di riva del litorale toscano e criteri di lettura. *Coste Toscan. Reg. Toscan.* **1989**, *2*, 33–56.
43. Anfuso, G.; Bowman, D.; Danese, C.; Pranzini, E. Transect based analysis versus area based analysis to quantify shoreline displacement: Spatial resolution issues. *Environ. Monit. Assess.* **2016**, *188*, 568. [CrossRef]
44. Pranzini, E. Bilancio sedimentario ed evoluzione storica delle spiagge. *Il Quaternario* **1994**, *7*, 197–202.
45. Cavazza, S. Regionalizzazione geomorfologica del trasporto solido in sospensione dei corsi d'acqua tra il Magra e L'Ombrone. *Alli Soc. Tosc. Sci. Nat. Mem.* **1984**, *91*, 119–132.
46. Becchi, I.; Paris, E. Il corso dell'Arno e la sua evoluzione storica. *Acqua Aria* **1989**, *6*, 645–652.
47. Bartolini, C.; Pranzini, E. Fan delta erosion in southern Tuscany as evaluated from hydrographic surveys of 1883 and the late'70. *Mar. Geol.* **1984**, *62*, 181–187. [CrossRef]
48. Innocenti, L.; Pranzini, E. Geomorphological evolution and sedimentology of the Ombrone River delta (Italy). *J. Coast. Res.* **1993**, *9*, 481–493.

49. Cappucci, S.; Scarcella, D.; Rossi, L.; Taramelli, A. Integrated coastal zone management at Marina di Carrara Harbor: Sediment management and policy making. *Ocean Coast. Manag.* **2011**, *54*, 277–289. [CrossRef]
50. Bowman, D.; Pranzini, E. Reversed response within a segmented detached breakwater—The Gombo case, Tuscany coast, Italy. *Coast. Eng.* **2003**, *49*, 263–274. [CrossRef]
51. Cipriani, L.E.; Pranzini, E. Differenziazione granulometrica dei sedimenti di spiaggia emersa e sommersa quale conseguenza della costruzione di opere a mare alla foce del fiume Cecina (Toscana centrale). *Boll. Soc. Geol. Ital.* **1990**, *109*, 471–480.
52. Anfuso, G.; Martínez, J.A.; Rangel Buitrago, N. Bad practice in erosion management: The southern Sicily case study. In *Pitfalls of Shoreline Stabilization*; Pilkey, O., Cooper, J.A.G., Eds.; Springer: Berlin/Heidelberg, Germany, 2012; p. 333.
53. Gandolfi, G.; Paganelli, L. Le province petrografiche del litorale toscano. *Boll. Soc. Geol. Ital.* **1977**, *96*, 653–663.
54. BEACHMED-e. Il Sottoprogetto OpTIMA, Ottimizzazione delle Tecniche Integrate di Monitoraggio Applicate ai Litorali. In *La Gestione Strategica Della Difesa Dei Litorali Per Uno Sviluppo Sostenibile Delle Zone Costiere Del Mediterraneo*; BEACHMED-e, I Quaderno Tecnico: Florence, Italy, 2007; pp. 33–47.
55. De Filippi, G.L.; Duchini, E.; Pranzini, E. Closure Depth estimation along the Tuscan coast aimed at short- and long-term coastal monitoring. In *Beach Erosion Monitoring-Results from BeachMed-e/OpTIMAL Project*; Pranzini, E., Wetzel, L., Eds.; Nuova Grafica Fiorentina: Firenze, Italy, 2008; pp. 33–50.
56. Hallermaier, R.J. Use for a calculated limit depth to beach erosion. In Proceedings of the XVI Coastal Engineering Conference, Hamburg, Germany, 27 August–3 September 1978; Volume 88, pp. 1493–1512.
57. Stive, M.J.F.; de Vriend, H.J. Modelling shoreface profile evolution. *Mar. Geol.* **1995**, *126*, 235–248. [CrossRef]
58. Bigongiari, N.; Cipriani, L.E.; Pranzini, E.; Renzi, M.; Vitale, G. Assessing shelf aggregate environmental compatibility and suitability for beach nourishment: A case study for Tuscany (Italy). *Mar. Pollut. Bull.* **2015**, *93*, 183–193. [CrossRef]
59. Bruun, P. Navigation and Sand Bypassing at Inlets: Technical Management and Cost Aspects. *J. Coast. Res.* **1996**, *23*, 113–119.
60. Nordstrom, K.F. *Beaches and Developed Coasts*; Cambridge University Press: Cambridge, UK, 2000; p. 338.
61. Khalil, S.M.; Finkl, C.W.; Roberts, H.H.; Raynie, R.C. New Approaches to Sediment Management on the Inner Continental Shelf Offshore Coastal Louisiana. *J. Coast. Res.* **2010**, *4*, 591–604. [CrossRef]
62. Boswood, P.K.; Voisey, C.J.; Victory, S.J.; Robinson, D.A.; Dyson, A.R.; Lawson, S.R. Beach Response to Tweed River Entrance Sand Bypassing Operations. In Proceedings of the 17th Australasian Conf. on Coastal and Ocean Eng. (incorporating 10th Australian Conf. on Port Eng.), Engineers Australia, Adelaide, Australia, 20–23 September 2005; pp. 119–124.
63. Pinto, C.A.; Silveira, T.M.; Teixeira, S.B. *Beach Nourishment Practice in Mainland Portugal (1950–2017): Overview and Retrospective*; Technical Report; Portuguese Environment Agency: Amadora, Portugal, 2018; p. 57.
64. Mason, M.A. *Abrasion of Beach Sand*; U.S. Beach Erosion Board, War Department: Washington, DC, USA, 1942; p. 26.
65. Chen, B. What Is the Contribution of Abrasion to the Sediment Budget of a Mixed Sand and Gravel Beach? Ph.D. Thesis, University of Otago, Dunedin, New Zealand, 2015.
66. De Falco, G.; Simeone, S.; Baroli, M. Management of beach-cast Posidonia oceanicaseagrass on the island of Sardinia (Italy, Western Mediterranean). *J. Coast. Res.* **2008**, *24*, 69–75. [CrossRef]
67. Cooper, J.A.G.; Anfuso, G.; Del Rio, L. Bad beach management: European perspectives. *Geol. Soc. Am.* **2009**, *460*, 167–179.
68. Anfuso, G.; Martinez, J.A.; Rangel, N. Morphological cells in the Ragusa littoral (Sicily, Italy). *J. Coast. Conserv.* **2013**, *17*, 369–377. [CrossRef]
69. Williams, A.T.; Micallef, A. *Beach Management: Principles and Practice*; Earthscan: London, UK, 2009.
70. Pranzini, E. Caratteristiche morfologiche e sedimentologiche di una zona di convergenza del trasporto litoraneo (Versilia, Toscana). *Studi Costieri* **2004**, *8*, 135–149.

MDPI
St. Alban-Anlage 66
4052 Basel
Switzerland
Tel. +41 61 683 77 34
Fax +41 61 302 89 18
www.mdpi.com

Journal of Marine Science and Engineering Editorial Office
E-mail: jmse@mdpi.com
www.mdpi.com/journal/jmse

www.ingramcontent.com/pod-product-compliance
Lightning Source LLC
Chambersburg PA
CBHW061618210926
51850CB00009B/2358
* 9 7 8 3 0 3 6 5 1 6 0 6 6 *